# Home Automation Basics II

## The LiteTouch® System

by James Van Laarhoven

# Home Automation Basics II

## The LiteTouch® System

by James Van Laarhoven

PROMPT® PUBLICATIONS

**PROMPT® Publications** is an imprint of Sams Technical Publishing, 5436 W. 78th St., Indianapolis, IN 46268.



International Standard Book Number: 0-7906-1226-7

Library of Congress Catalog Card Number: 00-105990

***Acquisitions Editor***: Alice J. Tripp
***Editor***: Kim Heusel
***Assistant Editor***: Will Gurdian
***Typesetting***: Kim Heusel
***Proofreader***: Pat Brady, Cricket Franklin
***Cover Design***: Christy Pierce
***Graphics Conversion***: Christy Pierce
***Illustrations and Other Materials***: Courtesy LiteTouch®



PRINTED IN THE UNITED STATES OF AMERICA

9 8 7 6 5 4 3 2 1

# Acknowledgments

Many facets of this book were completed with the cooperation and advice of these companies, corporations, and individuals: From LiteTouch, I would like to thank Walter Muller, Angie Larson and the entire technical support staff for their contributions and patient help during the writing of this book. I would like to thank Conrad Persson and Bob Rose for their helpful input. Appreciation and thanks for practical advice go to Joe and Liz Murphy from Coast to Coast Dimming and Control (LiteTouch dealers). I would also like to thank Kathy Schreier from GCI Dragonette. Special thanks to Larry and Sandy Berberich, who have given their kind advice and practical points of view after living with a LiteTouch automation system for so many years.

LiteTouch has contributed most of the photos and diagrams for this book. In addition, LiteTouch also supplied equipment specifications and training materials. Without them, the book would have been nearly impossible to write. Special thanks to LiteTouch for its generosity.

I would also like to give special thanks to Alice Tripp, Kim Heusel, and all of the editors at Prompt Publications for their guidance and hard work. Special thanks to the entire staff at Prompt Publications® and Sams Technical Publishing. If anyone has been excluded from this list, it was purely unintentional.

# Contents

# Preface

A home automation system generally covers a broad spectrum of household equipment control. It commonly includes systems such as lighting, environmental, entertainment, and security. The purpose of this book is to illustrate the practical steps involved in the installation of a home automation system. In addition, it will cover troubleshooting techniques. The text is meant to be an informative and interesting way for technicians/electricians, as well as homeowners, to meet the challenges of home automation. The widespread use of these systems seems inevitable; therefore, becoming familiar with digital equipment control will ensure your place in the home automation future.

LiteTouch manufactures the home automation system illustrated in this book. Starting with a basic design, the installation will progress until a completed automation system is both initiated and operable.

LiteTouch has been in the business of developing and improving home automation equipment and techniques for more than 20 years. Its name stems from origins that only a "light touch" was needed to activate its equipment. In 1988, LiteTouch introduced a true computer control system that was called LiteTouch 2000®. With this new computer control, programs could perform complex tasks automatically. Large, unsightly switch banks were eliminated and bulky high-voltage control wiring was reduced with this system. Safer low-voltage control wires and switches replaced the old standard wiring and controls that most of us are familiar with today. The LiteTouch Standard 2000 is still being used, but with many additional improvements from those earlier days. This system is installed in more than 10,000 buildings and the number is growing. The improved LiteTouch Standard 2000 along with LiteTouch's 5000LC will be the two home automation systems that will be covered in detail in this book.

The LiteTouch Standard 2000 system uses a Central Control Unit (CCU) as its main system control component. This computer (Figure P-1) receives digital signals from the switches (Figure P-2), then processes those commands and transmits the control data assigned to those

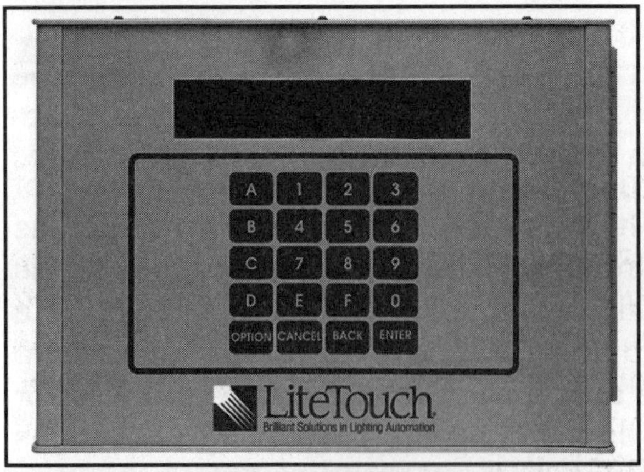


*Figure P-1*

switches to the control modules (Figure P-3). The load is then activated by the parameters that are programmed into the CCU's switch and load programs. The minimum equipment needed for a fully operational LiteTouch system is one CCU, one switch station, one control module, and one power supply (systems will always be larger than this).

Since this system mainly works from a digital base, information that passes within its environment tends to be more accurate than that of most analog home automation systems. Increased accuracy means fewer erroneous load activations and quicker response times throughout the automation environment. Button presses and external activations are therefore more immediate when using a digital system, so it tends to be the home automation choice when high quality is a concern.

There are four CCUs currently being used: the Standard, the Compact, the Elite, and the 5000LC. Each of these computers offers a unique capability for a specific application. The Standard CCU seems to be the most popular because of its versatility and dependability. The 5000LC is

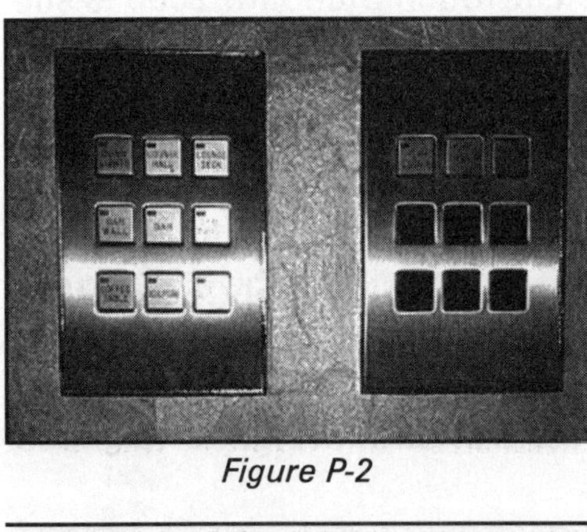

*Figure P-2*

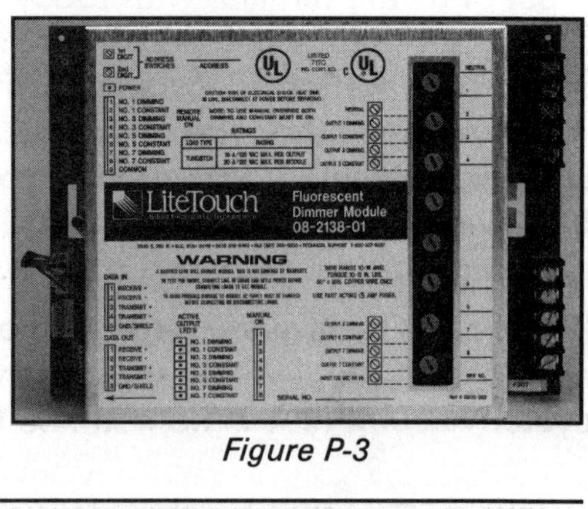


*Figure P-3*

a favorable addition because of its expansion capabilities and user-friendly programming environment. The same control switches and modules are compatible with all of these computers, allowing CCU upgrades to be accomplished with relative ease. More details about these computers will be discussed later in the text.

Switches (also called control stations or switch stations) can be activated manually by pressing a button or automatically by a timed function. Additionally, switches can be activated by interfacing inputs from equipment such as telephones, motion sensors, and photocells. Virtually any electrical or electronic device can activate CCU switch and load functions when properly interfaced.

The switches are addressed with a two-digit code for CCU identification. This code is dialed into the switch station by using two separate selectors. The switches themselves are wired from the CCU using a four-wire, 16-gauge unshielded cable. Only three of these wires are actually used in a typical switch system, however, there is a possible use for the fourth wire and that will be discussed in the troubleshooting chapter. The cables are daisy-chained through a series of electrical switch boxes and connected in parallel to the CCU switch station port.

Control modules (load controllers) are also addressed using a two-digit code, very much like that of the switch stations. The control module's purpose is to physically switch all of the high-voltage loads in the LiteTouch system (in certain circumstances, low-voltage loads can also be switched). A 22-gauge, two-twisted-pair shielded communications cable is run from the CCU to the enclosure that houses these modules. Control modules are available in many different configurations (dimmer, relay, and dc motor control, to name a few). The diversity of these and other control modules will be covered throughout the book.

Specific high-voltage connections and their applications will be discussed, in addition to transient surge protection that is required on these LiteTouch systems. Keeping high-voltage interference to a minimum is another important aspect of this system, and it will also be addressed.

To fully utilize the potential of a home automation system, control to and from non-system equipment is necessary. Interfacing from non-system equipment is sometimes accomplished through the use of a unit that is called the Data Input module. This module allows momentary or maintained contact closures to enable CCU programs to be initiated. Security systems can offer a dry-contact closure when tripped, thus enabling a security-type program that allows home equipment to be activated. For example, the entire lighting system can be programmed to turn on or flash upon an input activation. In addition, shutters and entry locks can also close and lock. At any time before or after the system is installed, changes can be made to the programming to add items that are found useful to security.

In one actual installation, the homeowners used a Data Input module to activate all of their house lights with a radio transmitter. A three-button transmitter supplied by the garage door company along with a receiver unit that closed a dry set of contacts accomplished this quite easily. Two of the buttons located on the remote activated the two garage doors and the third closed the separate set of dry contacts on the receiver, thereby activating the Data Input module. When the owners arrive home late, they press the third button on the transmitter and their house appears out of the dark. The Data Input module has many interesting interfacing applications.

The CCU telephone interface has several uses. With an optional modem installed, programming changes can be made off-location by technicians and programmers. This off-site interfacing capability can save the client and service people some valuable time. The CCU telephone interface can also be used to activate lighting loads or other equipment from anywhere in the world. For example, if you are visiting Madrid, Spain, and you hear that the weather is going to deteriorate where your home is in the U.S., just call your LiteTouch system and lower the shutters. In another example, you may need to let someone enter your house, but you cannot be there in person to do it yourself. In this type of situation, just call on the phone, unlock the front doors and turn on some needed lights. In one installation, the home automation system was coupled to the hot tub. The owner, who put in unusually long workdays, liked to use the hot tub when he arrived home. Furthermore, he did not like to wait the time it took for the tub to heat up. He would call home an hour before he left his office and activate the hut tub so it was warm and ready when he arrived.

Another interface device that works well with this system is an infrared (IR) remote and detector. An IR detector can be installed in the number 9 button location on a standard switch station. Using a handheld IR remote control, switch activation can be accomplished from anywhere in the room that the switch and detector are located. This can be a handy device in some automation systems.

An astronomical clock is included with the later CCUs. This feature allows programming functions to be performed at exact daytime and nighttime hours. Timer functions are generally set to activate loads during theoretical periods of darkness. This feature eliminates guesswork by compensating for the home's geographical location and current daylight saving time to activate the loads more accurately.

Interfacing the LiteTouch system with the home environmental controls can be accomplished in many ways. In one installation, the HVAC company provided two sets of contacts located at the main thermostat. One set was dry and would activate when a certain temperature or time was reached and another set would activate a program in the thermostat when they were shunted together. A one-pair cable was run from the dry set to the Data Input module and another one-pair cable went from the shunt contacts at the thermostat to a low-voltage relay module. Using this hardware setup, many things can be accomplished with imaginative programming. Motorized dampers can be installed in the main ducts that are moved via an output from a LiteTouch module, and external inputs from LiteTouch can activate programs in the thermostat.

Entertainment systems also can interface with LiteTouch. One such interface permits a TV to lower from the ceiling when someone walks into the room by the use of heat sensors coupled to the Data Input module. With proper interfacing, this same scenario could be voice activated. There are so many possibilities they are limited only by the imagination. Home automation is an open arena where everyone with an idea is heard. If an idea is good, it probably will be considered.

The last two items that will be mentioned before the installation begins are proper safety procedures and required test equipment. High-voltage wiring requires many skills and may also require an electrical contractor's license due to local and national electrical codes. High-voltage wiring techniques will be included in this book, however, it is not expected that someone without the experience in this field undertake these tasks without a qualified licensed electrician to assist.

Some of the electronic equipment in this system is highly susceptible to transient spikes and electrostatic discharge (ESD). Proper static precautions should be taken when working with certain components. The use of grounding wristbands and static mats may be a good idea in areas that can accumulate excess positive or negative charges. Usually, a dry environment is conducive to static buildup.

LiteTouch systems are not normally test-equipment intensive. Usually, a good-quality, sensitive multimeter and the LiteTouch power bus monitor clip are the only mandatory test equipment needed (the LiteTouch power bus monitor clip attaches to the data-out port of a control module to show if data is present). Obtaining a personal computer for system programming and diagnostics is advisable, although it is not needed for every service-type application. In extreme cases, signal meters and oscilloscopes are used to diagnose system problems. However, when this special test equipment is needed, it is usually only operated by LiteTouch personnel.

# Home Automation Basics II

## The LiteTouch System®

# 1

# Beginning the LiteTouch 2000 Installation

## Blueprint Reading, Equipment Planning, and Placement

The first step in a home automation installation is the planning stage. Prior to construction, the proposed LiteTouch system needs to be integrated into the main electrical blueprints. All electrical loads and switches are denoted on the electrical print with special notations (Figure 1-1). In this design, triangles attached to numbered squares are used to denote the switch control stations and boxed three-number codes denote the loads. The100-series numbers usually denote the loads assigned to the first floor and the 200-series numbers generally denote the loads on the second floor. However, in this plan, some 200-series numbers are used in the single-floor plan. The reason for this is to show the definite separation between the two different load control modules that are designated for this installation.

Generally, homes that incorporate an automation system are much larger than the one that is represented in this floor-plan layout. Nonetheless, the same principles of this layout will apply to any size home. This system is scaled down to avoid confusion.

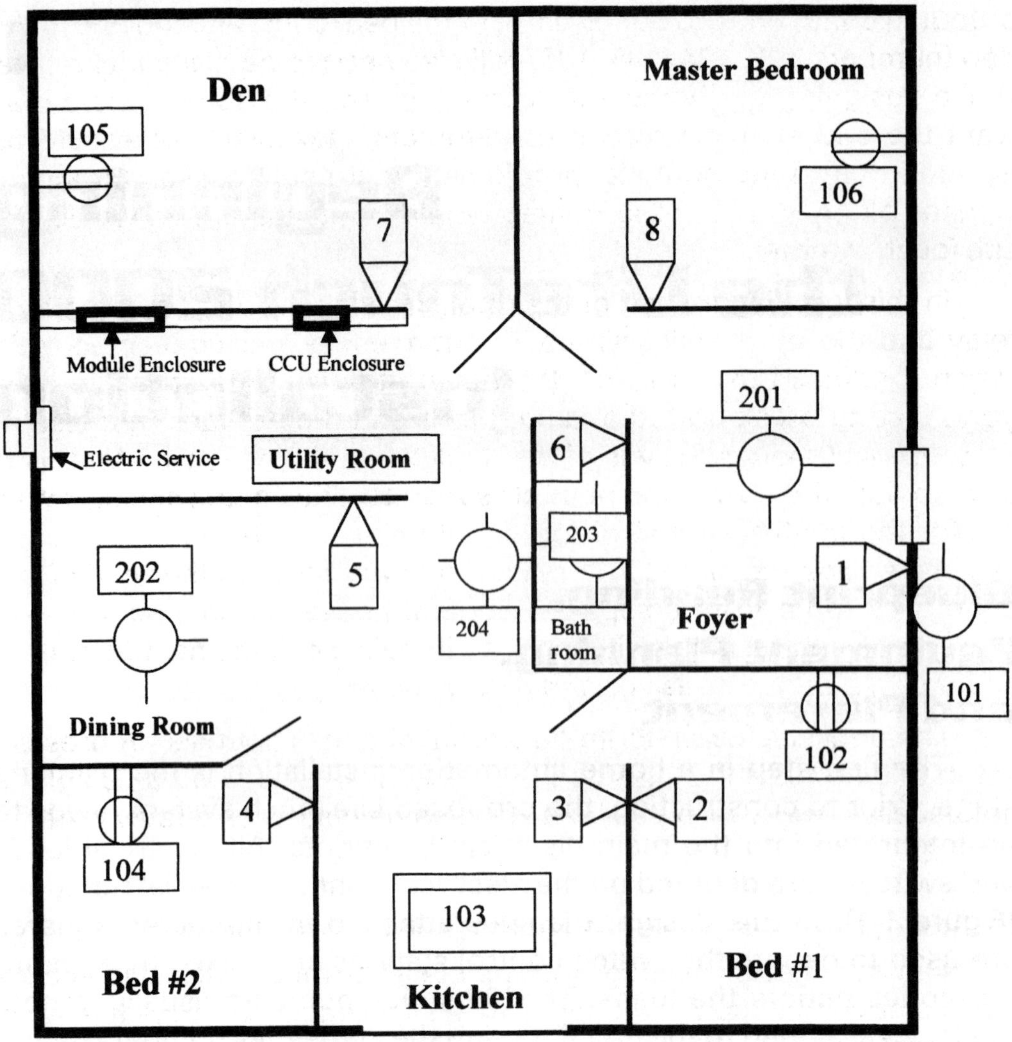


*Figure 1-1*

## Planning the Loads

The electrical loads that will be controlled by LiteTouch in this home are of two types: those that can be dimmed and those that cannot. The foyer, dining room, bathroom and hallway lights labeled 201, 202, 203, and 204, are incandescent and can be dimmed. The kitchen light will be a nondimming fluorescent type, so this load will need a relay module output to control it. The outlets in all of the bedrooms and the one in the den (numbers 102, 104, 105, 106) will also need to be placed on relays. While designing the home automation system, the homeowners may want the outlets to be placed on dimmers. However, this should be avoided due to the potential problems that it could cause. A vacuum cleaner plugged into these outlets would be damaged, as would the LiteTouch dimmer.

In this installation, half of the duplex outlet will be controlled by a relay and the other half will have permanent power connected to it. When it comes time to install the outlet at finish time, it would be a good idea to break both the neutral tabs and the hot tabs that electrically connect the two halves of the duplex outlet. The reason for this is to keep both the power and neutrals separate. The neutral and switched hot for the controlled half of the outlet will come from the LiteTouch module enclosure circuit. The neutral and hot for the other half of the outlet will come from a separate general house wiring circuit. If you shared one neutral in this situation, an imbalance of the neutral system could cause undue stress on the neutral wires.

The only LiteTouch-controlled circuit that has not been addressed yet in this home is the outside light (101). This type of outside light is not usually dimmed, so this load will also be placed on a relay module output.

### Planning for the Load Modules and Enclosures

The information needed to complete a LiteTouch enclosure schedule has now been compiled (Figure 1-2). The enclosure schedule is an important part of the planning stage in this system. When properly completed, it will illustrate where each individual load control will be located in the control module enclosure. It will also indicate wattage information and what kind or how many electrical distribution circuits will be needed. Module addresses and types are also stated on this enclosure schedule. The relay module 01 in this enclosure schedule controls four outlets, one fluorescent, one 60-watt fixture, and will be supplied by one 20-amp circuit. The wattage use of the outlets will be an educated guess. The reason for this is that the accurate power usage of an outlet is usually indeterminable. Most likely, some type of table or floor

## Control Module Enclosure Schedule

Module address: 01 Module type: Relay 6-channel No.of enclosures: 1

| MODULE POSITION | LOAD # | LOAD LOCATION | WATTAGE | CIRCUIT SUPPLY |
|---|---|---|---|---|
| 1 | 101 | Outside Wall Light | 100 | 1- 20 amp 110v circuit |
| 2 | 102 | Bed #1 Outlet | 200 | |
| 3 | 103 | Kitchen Light | 200 | |
| 4 | 104 | Bed #2 Outlet | 200 | |
| 5 | 105 | Den Outlet | 200 | |
| 6 | 106 | Master Bed Outlet | 200 | |
| | | Total Load: | 1100 | |

Module address: 02 Module type: Dimmer module

| POSITION | LOAD# | LOAD LOCATION | WATTAGE | CIRCUIT SUPPLY |
|---|---|---|---|---|
| 1 | 201 | Foyer Light | 200 | 1- 20 amp 110v circuit |
| 2 | 202 | Dining Light | 200 | |
| 3 | 203 | Bathroom Light | 200 | |
| 4 | 204 | Hallway Light | 100 | |
| | | Total Load: | 700 | |

| | |
|---|---|
| POSITION 3<br>POWER SUPPLY MODULE (SINGLE)<br>OUTPUT: dc | 1-20 AMP<br>110V CIRCUIT |
| POSITION 4<br>DATA INPUT MODULE (MOMENTARY)<br>CONTROL: Security Interface<br>Switch Station Address: FF-1 | N/A |

*Figure 1-2*

lamp will be used in these locations. The type of lamps to be used is not known; however, even with some fairly high-wattage lamps, the load should be low enough for the limitations of both the circuit and the LiteTouch relay module. The wattage figure that is used for each outlet will be figured at 200 watts and the total watts computed for the entire relay module is 1100 watts.

The dimmer loads 201, 202, 203, and 204 have a combined wattage requirement of 700 watts. This wattage is well within the limits of both the dimmer control module and the 20-amp circuit feeding it.

Beneath the dimmer control module in the enclosure schedule is the system power supply. This low-voltage power supply will be used to power the Standard 2000 CCU, the relay module, and the switch control stations. Beneath the power supply is the last module in this enclosure schedule, the Data Input module. The Data Input module in this system will be interfaced only with the security system. The wiring and connections for the entire enclosure will be explained further along. The enclosure schedule is finished for now.

Finding an area on the print that will contain the LiteTouch enclosures and CCU can sometimes be a challenge. Living spaces such as bedrooms, and dining and living rooms are usually not good ideas for many reasons. One of the main reasons is that it is hard to hide the enclosure panel because of its size. Sometimes pictures can be hung in front to hide it from view, but that leaves the equipment not readily accessible. It also may inhibit airflow to the panels, thereby decreasing the cooling requirements of the system. In addition, some of the control modules emit some sound and this could disturb the homeowners. Garages are sometimes acceptable, but it all depends on the surrounding temperature. The recommended ambient temperatures for the CCU are 45 degrees F (low) and 95 degrees F (high). The control module temperatures are 32 degrees F (low) and 95 degrees F (high). Many garages around the country surpass both the low and high recommendations at some time during the course of the year. Subjecting the equipment to extreme temperature variations can seriously affect the operation and life of the internal electronics. The best idea for the placement of this equipment is out of sight and in an environmentally controlled atmosphere. Indoor utility rooms are the most common choice for home automation equipment location sites.

Another consideration to keep in mind when planning the placement of equipment is accessibility. The door on a control module enclosure measures 39 inches high and 22 inches wide. Mounting the panel too close to other doors or cabinets could limit access. Consulting some of the other building trade people that will be working on the house may be helpful in finding just the right location. In this installation, the control module enclosure and the CCU will be placed in the utility room between the dining room and the den. It is important to note that in all LiteTouch installations, the CCU needs to be located within 50 feet of the enclosure that contains the power supply or power supplies that feed the CCU. This is not a concern in this particular installation, since both the CCU and the enclosure that will house the power supply will be located in close proximity to each other.

## Designing the Switch Configurations

A station design guide is used for preliminary switch control station planning (Figure 1-3). Later, this same guide can be used to install the switches in the house and also to program the switches into the CCU. During this part of the planning stage, the guide will permit you to gather accurate information for switch station ordering. Switch location, configuration, color, and button engraving are included, in addition to the load numbers and module addresses.

There are many switch types to choose from. For this installation, however, we will choose the S-Series-type switches. The S-Series switch accommodates up to nine buttons in a single-gang configuration and comes standard with an LED indicator located over each control button. Starting with switch number 01, we will go over the plans for each station design guide. Please refer to Figures 1-1 and 1-2 in relation to each station design guide figure. At the top of each station design guide, nine switch (or button) position configurations can be observed (Figure 1-3). Corresponding numbers are assigned to each individual button. These numbers represent part of an actual electronic switch address that the CCU will recognize. Switch station number 01 has a five-button configuration. Referencing the five-button figure at the top of the design guide, you will notice that button numbers 1, 3, 5, 7, and 9 are the designated electronic address extensions for switch 01. The CCU will recognize each one of these extensions by the prefix 01 placed before the button numbers, for example, 01-1, 01-3, 01-5, 01-7, and finally 01-9.

## Station Design Guide

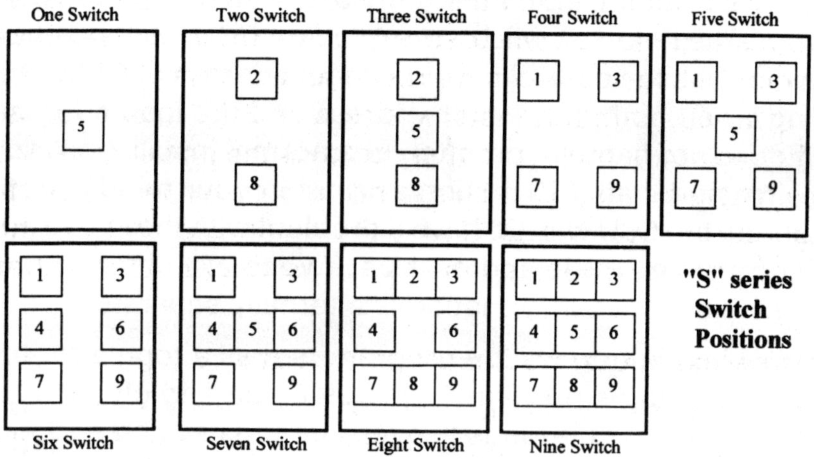


CONTROL STATION ID ADDRESS: **01** NUMBER OF BUTTONS: 5

CONTROL STATION LOCATION: **FOYER**

| CONTROL SWITCH POSITION:<br>Button color:<br>LED color: | 1 | 2 | 3 | 4 | 5 | 6 | 7 | 8 | 9 |
|---|---|---|---|---|---|---|---|---|---|
| Engraving schedule: | Outside Light | | Foyer | | Kitchen Path | | House off | | House on |
| Switch Type: | Toggle | | Dim | | Scene | | Toggle Off | | Toggle on |
| Load Number: | 101 | | 201 | | 103<br>201<br>204 | | All Loads | | All Loads |
| Module Address: | 01-1 | | 02-1 | | 01-3<br>02-1<br>02-4 | | 01-1<br>01-2<br>01-3<br>01-4<br>01-5<br>01-6<br>02-1<br>02-2<br>02-3<br>02-4 | | 01-1<br>01-2<br>01-3<br>01-4<br>01-5<br>01-6<br>02-1<br>02-2<br>02-3<br>02-4 |

*Figure 1-3*

Programming will not be covered in this chapter; however, it may help to mention some programming techniques in this planning section. The switch address 01-1 (Figure 1-3) will be designated as the "Outside Light." The load that will be assigned to this switch position is print #101, which has the control module (load) address of 01-1. This can be confusing when both the switch address and the load address are the same. It does not happen too often in a normal installation, so keeping most switch and load information straight in your mind is not difficult. Rest assured, the CCU will not have a problem with these two addresses. The CCU considers these addresses as two totally different entities because switch and load information is internally separated.

This switch button will be programmed as a toggle switch, which means it simply turns position 1 of relay 01 on and off as you depress the button. The next button is "Foyer." Its CCU switch address is 01-3 and will dim the address load 02-1 when that button is depressed and held in (the 201 load number is a print designation and plays no electronic control role in the CCU). As the switch button is held in, it will control the load throughout the whole dimming cycle. It will stay at a set consistent dimming level when you let go of the button. In the future, whenever you tap this button on, it will dim the load to the exact same level as you set earlier.

Switch button 01-5 is labeled "Kitchen Path." Depressing this button from the front door will turn on the foyer light load 02-1 (print #201), the hallway light load 02-4 (print #204), and the kitchen light load 01-3 (print #103). It is designated as a "Scene" because different percentage levels of light will be assigned to the dimming loads at programming time. Switch station 01-7 is labeled "House Off." This button will be programmed to toggle-off every LiteTouch load in the entire house. This is a great feature in the larger homes because you are assured that every light has been turned off. You will not have to worry that any lights are on before going to bed. Switch station 01-9 is labeled "House On." Depressing this button will do just the opposite of what switch 01-7 accomplishes—it will toggle-on every load in the house. This is a good security button. If you think you are hearing sounds from somewhere in the house at night, press this switch and the entire house will illuminate.

On to the next switch, which is located in Bedroom #1. Switch 02 (Figure 1-4) is a three-button station in Bedroom #1. Button 02-2, labeled "Room," is an on and off control of module address 01-2 (Print #102). The relay module controls half of a duplex outlet for a floor or table lamp. Button 02-5, labeled "Kitchen," is also an on and off control and sends a digital signal to the CCU to activate load 01-3 (print #103). The last button on this station is labeled "Bath Path," and it is a "Scene" type control. When depressed, it will activate address loads 02-3 and 02-4 (print #'s 203 and 204). The preset scene levels will then dim both loads to the percentage that was designated into the program of the CCU. For example, 02-3 could dim to 50% and 02-4 could dim to 40%. This button would most likely be activated at night when someone needs to use the bathroom. Keeping the lighting levels low may be a good idea for this type of application.

The next switch station is depicted in Figure 1-5. The kitchen switch station ID address is 03. It is a four-button switch. Therefore, the address information that is sent to the CCU will come from 03-1, 03-3, 03-7, or 03-9. When button 03-7 is depressed, the computer sends a digital signal to relay module 01 and tells it to turn on its third load to 100% (relays can only operate the load at 100%). In the same signal, it also tells dimmer module 02 to turn on its first and fourth loads to a certain percentage. What appear in the house are the kitchen light, the dimmed hallway light, and the dimmed foyer light. The person who pressed this button in the kitchen can now walk to the front door and answer it without turning on three different light switches. If that person was not answering the door, but actually leaving the house, he or she could press the "Kitchen Path" button in the foyer and turn all those lights off without having to go back to the kitchen.

Bedroom #2's control station ID address is 04 and it is a three-button control station (Figure 1-6). It resembles the configuration and labeling of switch number 02 in bedroom #1. However, switch station 04-2 has a different module address for the switched outlet. It is load address 01-4 (print #104). The switches in bedrooms #1 and #2 control the same kitchen and bath path loads.

The dining room in this plan has a three-button control station on the opposite wall of the utility room. The communications address for

## Station Design Guide

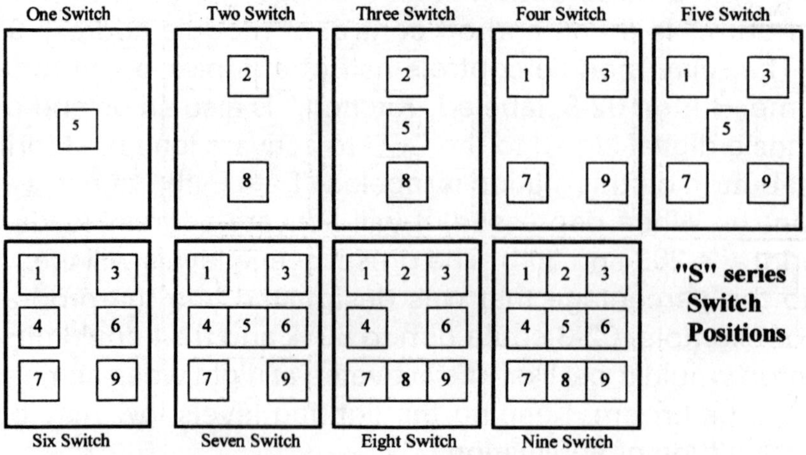


CONTROL STATION ID ADDRESS: **02** NUMBER OF BUTTONS: 3

CONTROL STATION LOCATION: **Bedroom #1**

| CONTROL SWITCH POSITION:<br>Button color:<br>LED color: | 1 | 2 | 3 | 4 | 5 | 6 | 7 | 8 | 9 |
|---|---|---|---|---|---|---|---|---|---|
| Engraving schedule: | | Room | | | Kitchen | | | Bath<br>Path | |
| Switch Type: | | Toggle | | | Toggle | | | Scene | |
| Load Number: | | 102 | | | 103 | | | 203<br>204 | |
| Module Address: | | 01-2 | | | 01-3 | | | 02-3<br>02-4 | |

*Figure 1-4*

## Station Design Guide

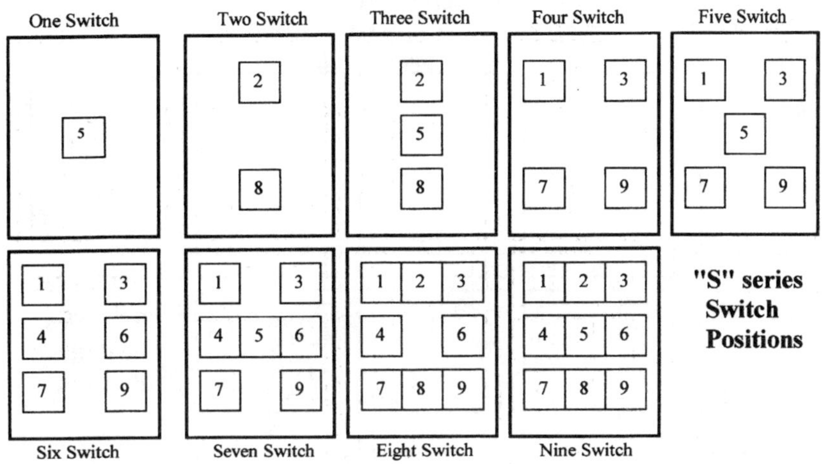


CONTROL STATION ID ADDRESS: **03** NUMBER OF BUTTONS: **4**

CONTROL STATION LOCATION: **Kitchen**

| CONTROL SWITCH POSITION:<br>Button color:<br>LED color: | 1 | 2 | 3 | 4 | 5 | 6 | 7 | 8 | 9 |
|---|---|---|---|---|---|---|---|---|---|
| Engraving schedule: | Kitchen | | Bed #1 | | | | Kitchen Path | | Master Bed Path |
| Switch Type: | Toggle | | Toggle | | | | Scene | | Scene |
| Load Number: | 103 | | 102 | | | | 103<br>201<br>204 | | 106<br>204 |
| Module Address: | 01-3 | | 01-2 | | | | 01-3<br>02-1<br>02-4 | | 01-6<br>02-4 |

*Figure 1-5*

Figure 1-6
Station Design Guide

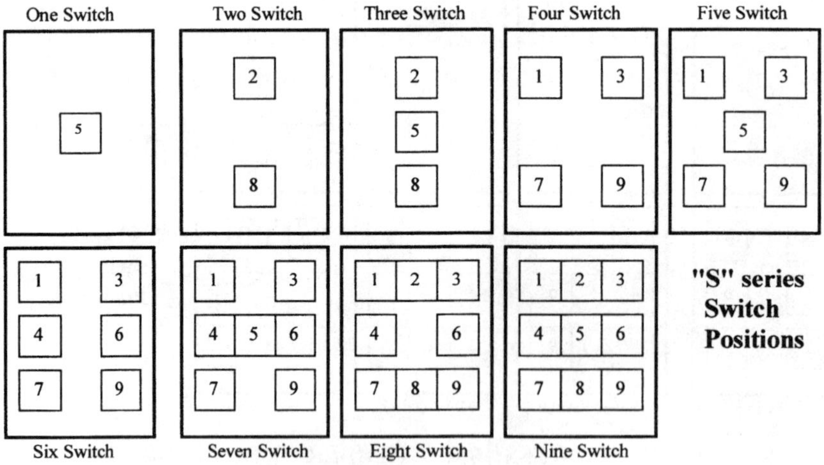


CONTROL STATION ID ADDRESS: **04** NUMBER OF BUTTONS: **3**

CONTROL STATION LOCATION: **Bedroom #2**

| CONTROL SWITCH POSITION:<br>Button color:<br>LED color: | 1 | 2 | 3 | 4 | 5 | 6 | 7 | 8 | 9 |
|---|---|---|---|---|---|---|---|---|---|
| Engraving schedule: | | Room | | | Kitchen | | | Bath<br>Path | |
| Switch Type: | | Toggle | | | Toggle | | | Scene | |
| Load Number: | | 104 | | | 103 | | | 203<br>204 | |
| Module Address: | | 01-4 | | | 01-3 | | | 02-3<br>02-4 | |

*Figure 1-6*

this switch is 05 (Figure 1-7). Button 05-2 is programmed as a dimmer and controls module address 02-2 (print #202). Button 05-5 (kitchen) is programmed as an on and off toggle switch that controls module address 01-3 (print #103). Button 05-8 is the last button on this switch and it controls the hallway light 02-4 (print #204). It should be noted that these programs can be changed quite easily, even after the system has been installed for years. Using button 05-8 as an example, we will see why the owners now want to change it. They have children who like to change the dimmer settings on the hallway light. The owners complain that the light is always too bright and that they are tired of resetting it back to the proper level. This can be remedied easily by changing the original programming of the switch to activate a scene program rather than a standard dimmer program. After the old programming for the button has been changed from a dimmer to a scene, you would then enter a desired preset dimming percentage into the scene program. The children could now press and hold the same button the way they had before, but the hallway light would always turn on to the same level that was programmed on the scene preset.

Switch station 06-5 (Figure 1-8) is located in the bathroom. It is a one-button switch station that controls only the bathroom light. The address for the control module load is 02-3 (print #203). If the owners of the house ever decide that they want this switch to have more buttons, they can order another switch in any configuration they desire. The new switch could be installed fairly quickly. It would be a small matter to change the programming at the CCU to accommodate for the additional buttons and their corresponding loads.

The den has a two-button switch station with an address of 07 (Figure 1-9). The two station ID addresses that this switch sends to the CCU are 07-2 and 07-8. The number 2 position on the switch is programmed as a toggle and controls load address 01-5 (print #105). The number 8 position on this switch is programmed as a dimmer and controls load address 02-4 (print #204).

The last control station that needs to be covered is address 08 (this is not including the Data Input which will be covered later on in the book). Switch station 08 (Figure 1-10) is located in the master bedroom. Generally, master bedrooms require a greater number of automation

## Station Design Guide

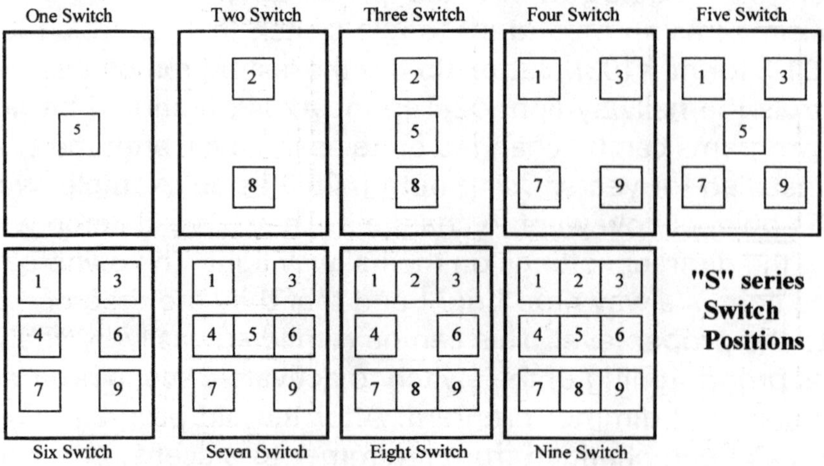


CONTROL STATION ID ADDRESS: **05** NUMBER OF BUTTONS: 3

CONTROL STATION LOCATION: **Dining Room**

| CONTROL SWITCH POSITION:<br>Button color:<br>LED color: | 1 | 2 | 3 | 4 | 5 | 6 | 7 | 8 | 9 |
|---|---|---|---|---|---|---|---|---|---|
| Engraving schedule: | | Dining Room | | | Kitchen | | | Hall way | |
| Switch Type: | | Dim | | | Toggle | | | Dim | |
| Load Number: | | 202 | | | 103 | | | 204 | |
| Module Address: | | 02-2 | | | 01-3 | | | 02-4 | |

*Figure 1-7*

## Station Design Guide

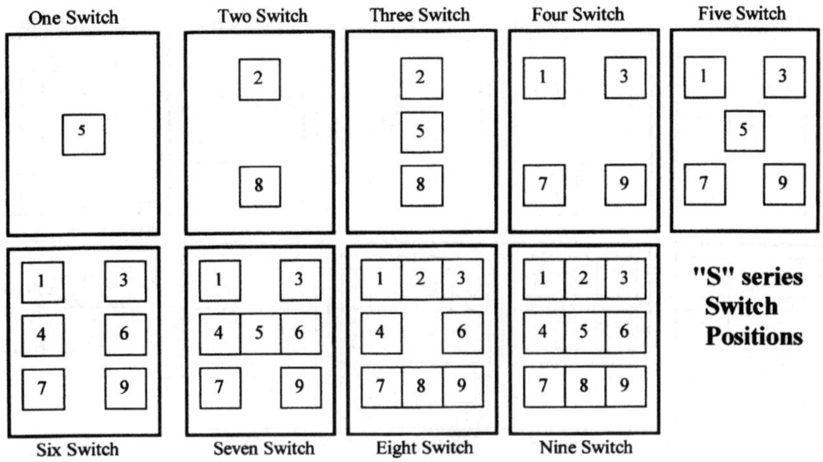


CONTROL STATION ID ADDRESS: **06** NUMBER OF BUTTONS: **1**

CONTROL STATION LOCATION: **Bathroom**

| CONTROL SWITCH POSITION:<br>Button color:<br>LED color: | 1 | 2 | 3 | 4 | 5 | 6 | 7 | 8 | 9 |
|---|---|---|---|---|---|---|---|---|---|
| Engraving schedule: | | | | | Bath | | | | |
| Switch Type: | | | | | Dim | | | | |
| Load Number: | | | | | 203 | | | | |
| Module Address: | | | | | 02-3 | | | | |

*Figure 1-8*

## Station Design Guide

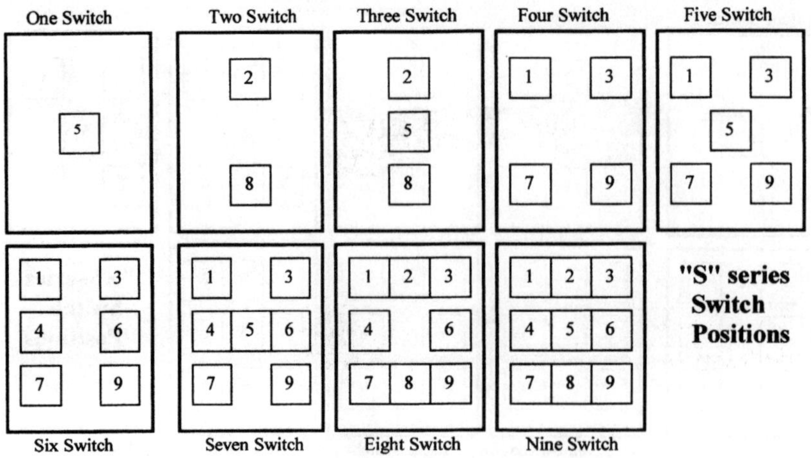


CONTROL STATION ID ADDRESS: **07** NUMBER OF BUTTONS: **2**

CONTROL STATION LOCATION: **Den**

| CONTROL SWITCH POSITION:<br>Button color:<br>LED color: | 1 | 2 | 3 | 4 | 5 | 6 | 7 | 8 | 9 |
|---|---|---|---|---|---|---|---|---|---|
| Engraving schedule: | | Den | | | | | | Hall way | |
| Switch Type: | | Toggle | | | | | | Dim | |
| Load Number: | | 105 | | | | | | 204 | |
| Module Address: | | 01-5 | | | | | | 02-4 | |

*Figure 1-9*

controls as compared with other rooms in the home. The owners usually occupy this bedroom, so it does make sense. This is a six-button control station and the number 1 button controls the half-switched outlet load 01-6 (print # 106). The number 3 button is a dimmer programmed to activate load 02-4 (print #204). Switch 08-4 is the kitchen path programmed as a scene. Loads 01-3 and 02-4 (print #'s 103 and 204) are activated by this switch to allow adequate lighting to the kitchen from the master bedroom. Switch 08-6 controls the dimming load in the bathroom that is designated as load address 02-3 (print #203). Switch 08-7 turns off all of the loads in the house when depressed. Switch 08-9 turns on all of the loads in the house when depressed. The planning stage of the LiteTouch system is now complete for this house. The next phase is the rough-in part of the install.

## Rough-In Stage

Now that the preliminary planning has been completed and the house is now in the framework stage, the rough-in phase can begin. This step in the building process is called "rough-in" because it is in the rough construction phase and most tradesmen need to put in material that is essential to the completion of the final phase. In this stage, the Standard 2000 CCU and control module enclosures need to be installed first (Figure 1-11). These are empty cans and will not have the electronic equipment installed in them until the house is nearly complete. Referring to Figure 1-12, the two enclosures need to be mounted inside the stud wall in the utility room. Half-inch drywall will be used over the studs, so make sure that the enclosures protrude out from the flush stud a shade less than half an inch. Make sure both enclosures are level and screwed into the studs securely. When this is accomplished, we will reference back to Figure 1-12 to install the switch station boxes.

Plastic or phenolic electrical switch boxes are very common in residential wiring, so this is the type that will house the LiteTouch switch stations. Make sure that the boxes are deep enough (about 3 inches) and that the opening is large enough to accommodate the switch station. All single-gang boxes are not created equal. Even though most standard switches and outlets will fit in a single-gang box, LiteTouch switches require more room. Just to be safe, use the largest single-gang box that is available. The next step is to find out the specified

Station Design Guide

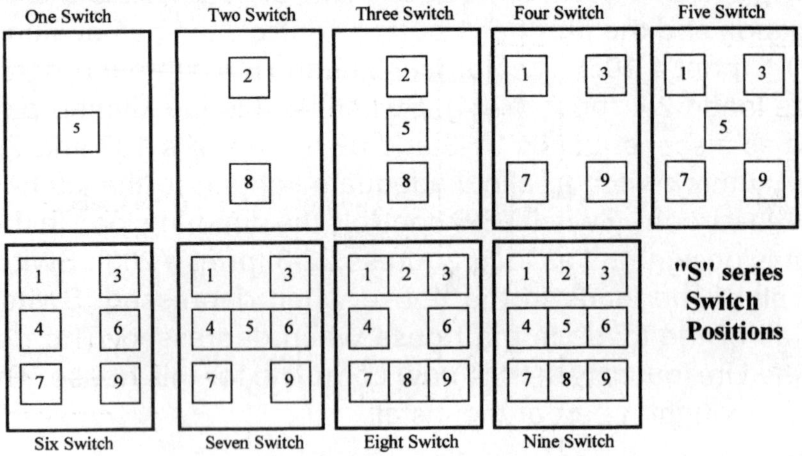


CONTROL STATION ID ADDRESS: **08** NUMBER OF BUTTONS: **6**

CONTROL STATION LOCATION: **Master Bedroom**

| CONTROL SWITCH POSITION:<br>Button color:<br>LED color: | 1 | 2 | 3 | 4 | 5 | 6 | 7 | 8 | 9 |
|---|---|---|---|---|---|---|---|---|---|
| Engraving schedule: | Room | | Hall way | Kitchen Path | | Bath | House off | | House on |
| Switch Type: | Toggle | | Dim | Scene | | Dim | Toggle Off | | Toggle on |
| Load Number: | 106 | | 204 | 103<br>204 | | 203 | All Loads | | All Loads |
| Module Address: | 01-6 | | 02-4 | 01-3<br>02-4 | | 02-3 | 01-1<br>01-2<br>01-3<br>01-4<br>01-5<br>01-6<br>02-1<br>02-2<br>02-3<br>02-4 | | 01-1<br>01-2<br>01-3<br>01-4<br>01-5<br>01-6<br>02-1<br>02-2<br>02-3<br>02-4 |

*Figure 1-10*

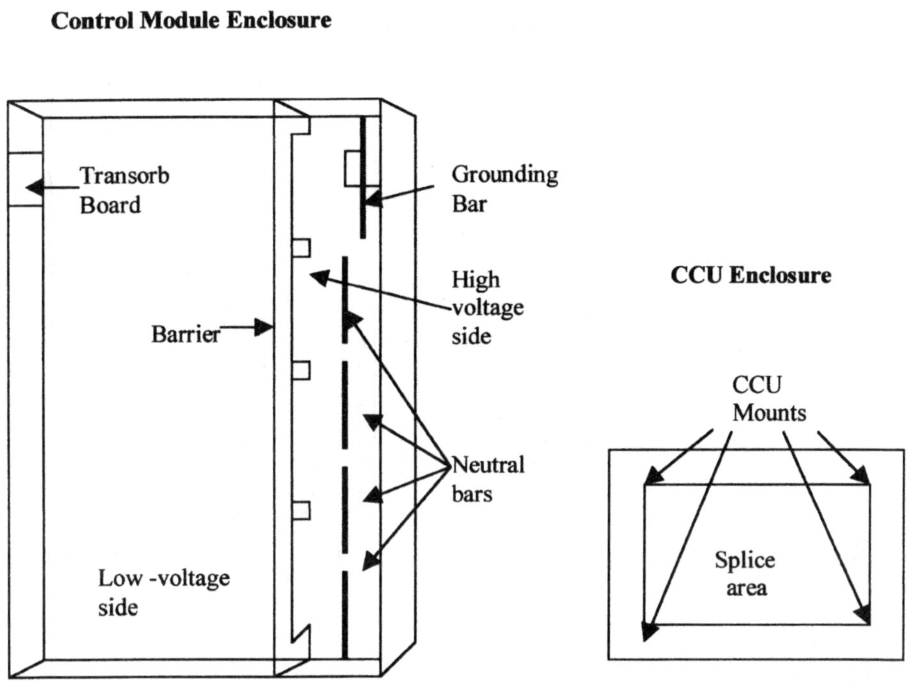


*Figure 1-11*

switch height for the house. About 44 inches to the bottom of the box is common, but ask to make sure. Attach each single-gang box referencing their locations on the print. Remember to let the box protrude from the flush stud a little less than a half-inch. Also, be sure to check that the entry doorway swings will not interfere with the switch boxes that you just installed. In this home, there should be eight LiteTouch switch control station boxes installed.

## Low-Voltage Wiring

With the enclosures and switch boxes installed, it is now time to install the low-voltage wiring runs. In preparation for both the low- and high-voltage cable runs, holes must be drilled in the top framing plate above each box and above both enclosures. One three-quarter-inch hole above each low-voltage switch box should suffice and two-three-quarter-inch holes in the top plate above the CCU enclosure. Several such holes should be drilled above the control module enclosure to accom-

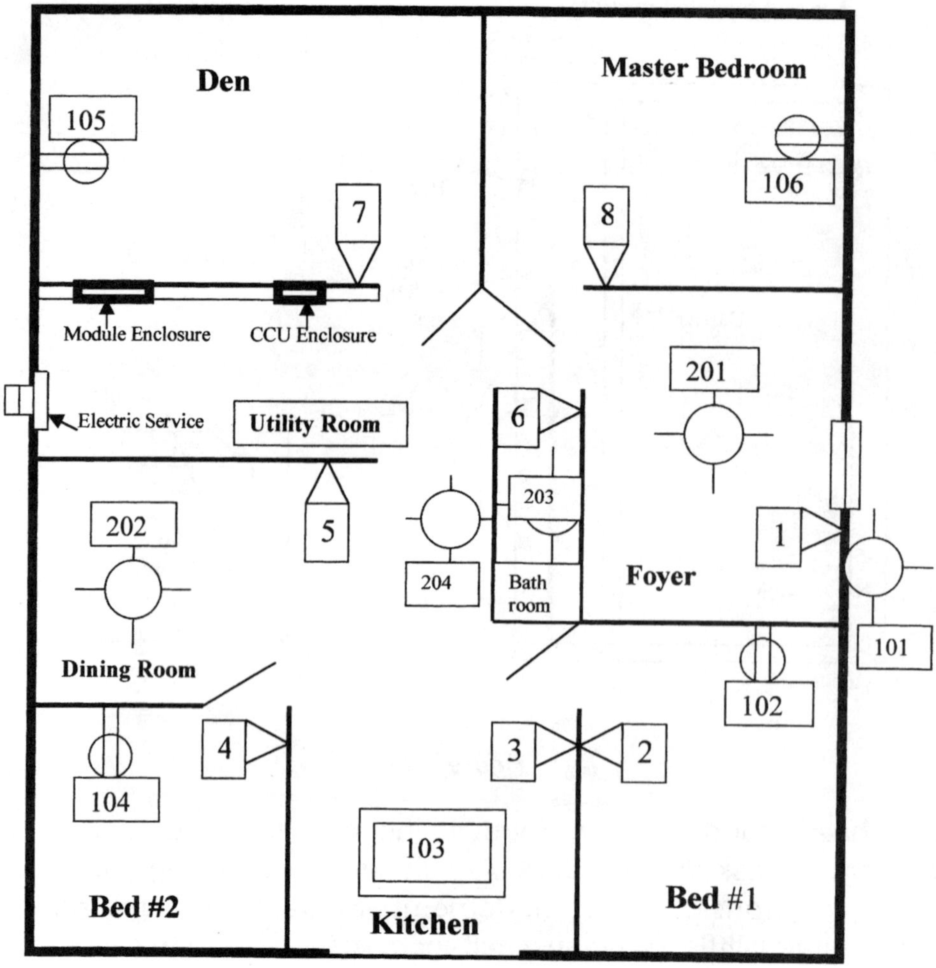


*Figure 1-12*

modate both the low- and high-voltage cables. The reason the holes should be drilled before any cables are run is to reduce the chance that the drill bit may damage any cables that may be out of view to the driller. Keep the low-voltage runs in separate holes from the high-voltage. Also, try to keep the low-voltage cables as far away from the highs as possible throughout the house to reduce interference. LiteTouch recommends at least a 6-inch clearance.

The control module enclosure is partitioned into two areas. The left side is for low-voltage and the right side is for high-voltage only. Run the low-voltage wires as far left in the enclosure as you can, so as to not interfere with the future control module mounting area. Use cable connectors where you penetrate the CCU and control module enclosures to reduce any damage that may occur to the low-voltage cables. When you run the high-voltage cables later on, the use of cable connectors is required by code.

The first cable that will be run is the control module communication wire. This cable is a 22-gauge, two-twisted-pair shielded assembly (LiteTouch No.08-2105, Belden No.9502, or equivalent). Run a single cable from the CCU enclosure into the control module enclosure. Leave enough cable inside the module enclosure to reach the top of the left-hand corner (near the Transorb board). Also, leave about 2 feet of cable inside the CCU enclosure to ensure enough cable for the future CCU attachment. In larger systems with more than one control module enclosure, you would daisy-chain this communication cable from module enclosure to module enclosure. It is acceptable to run another communication cable out of the CCU enclosure to each separate module enclosure. In some instances, this may be a very good idea.

Next, using an unshielded four-conductor 16-gauge cable (LiteTouch No.08-2103, Belden No.8465, or equivalent), run it from the CCU to the control module enclosure, just like you did for the communication cable (this unshielded wire will bring dc power from the power supply in the module enclosure to the CCU). Mark the end of the cable "From power supply" in the CCU enclosure. Mark the other end of the cable "Power to CCU." Using the same type of unshielded cable, run from the CCU enclosure to the control module enclosure one more time (this cable is the control station connection for the Data Input module). Mark the end of the cable that you ran into the module enclosure, "Control Station for the Data Input module." Mark the other end of the cable that is in the CCU enclosure, " To Data Input control station."

Next, it is time to wire the eight house switch control stations. Starting from the CCU enclosure, install a four-conductor 16-gauge unshielded cable to the nearest of the LiteTouch switch boxes (this would be switch station 7 in the den). Leave about 6 inches of cable protruding out past the front edge of the switch box. Back at the CCU enclosure, mark this cable, "Switch station run." Leave about 2 feet of cable inside the CCU enclosure. Continue looping the control station cable (continuing from switch station 7) into each LiteTouch switch station box until all eight are picked up. Staple the cables securely so they will not get in the way of the drywall when it is installed. Tuck the two cables back into all of the single-gang boxes (the last switch box you ran the unshielded cable to will only have one cable in it). You will splice these when you come back for the electronics installation on the finish.

### Installing the Telephone Cable

Using a two- or three-pair telephone cable, run from the CCU to the outside telephone interface box. This should be a dedicated run. Mark the cable in the telephone interface box to show that it is for the LiteTouch CCU. These connections will be terminated when the electronics are installed on finish.

### Security Wiring

A meeting with the security company is a good idea before you install any low-voltage cabling to its system. Most of the time, security people are very glad to accommodate your interfacing requests. Generally, you only need to ask them for an isolated single-pole dry contact closure for you to connect to. There is usually plenty of space available in their security box for them to mount a small activating relay with two contacts. Make sure you find out if the security output relay will be maintained or momentary, since the proper Data Input module needs to be ordered. Many times, the security people will make the connections inside their box for you if you give them some instruction. The security box is usually locked and may not be accessible when you arrive to do the final installation.

In most cases, the unshielded cable that you used for the control stations will work well for the security interface. The only time this is not true is when a more complex security interface is used. Shielded cables may need to be used in this type of situation. Complex systems will usually not sneak up on you. When dealing with more complex systems, the security people and the installer are usually more informed before the project even starts, so this should not be a great concern.

Low-voltage cabling

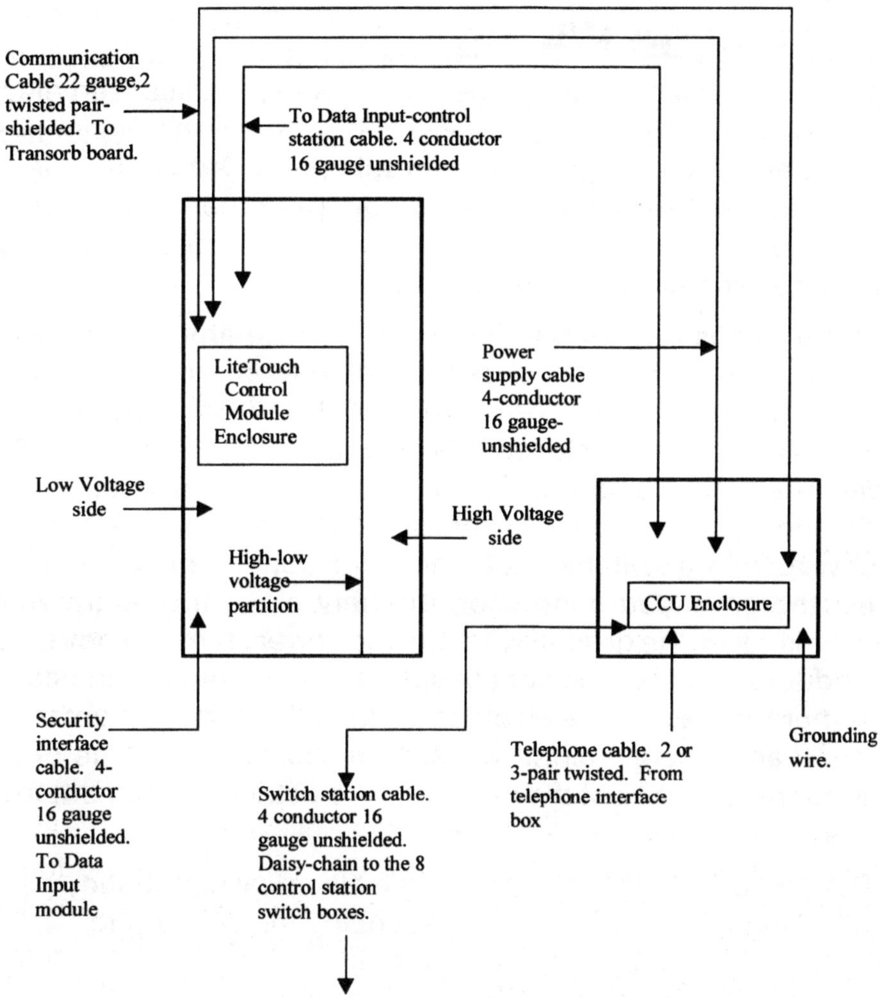


*Figure 1-13*

In the simpler cases, run this four-conductor cable from the security control box to the Data Input module. The final Data Input connection will be made at finish time. Make sure you keep the low-voltage wiring to the left of the control module enclosure partition. The right side of the internal partition is for high-voltage only. Also, use cable connectors in the knock-out areas of the enclosure to ensure that the cables are protected from the sharp metal edges. Figure 1-13 depicts the complete low-voltage cabling diagram for this installation.

## High-Voltage Wiring

Most local building departments require that a licensed electrician perform all of the high-voltage wiring and connections. National electrical codes also have requirements in this area. Ordinarily, it is not expected for the LiteTouch installer to complete the high-voltage work, but it will be covered here so that the installer is familiar with the proper installation techniques.

In the enclosure schedule, three different 20-amp circuits are specified. 12-2 nonmetallic (NM) sheathed cable (commonly called Romex) has three wires located beneath the outer covering: one enclosed in black insulation, one enclosed in white insulation, and the other bare copper. The black conductor is called by these names and others depending on how it is used: the ungrounded conductor, the hot wire, the power wire, or the switched hot. The white conductor is called by these names: the grounded conductor, the neutral, or the return wire. The bare wire is called the ground wire, but the correct terminology is grounding conductor. This wire is not considered a current-carrying conductor unless there is a fault in the electrical system. It will then shunt the power to ground and cause a circuit breaker or fuse to blow. The grounding wire is connected to all of the metal of the electrical equipment for safety reasons.

Run three 12-2 NM cables from the electrical distribution panel into the right side of the control module enclosure (be sure to use Romex connectors). Mark the outer sheathing on one of the cables, "LiteTouch relay module" at both ends. Mark one of the other cables, "LiteTouch dimmer module" at both ends. Finally, mark the last cable, "LiteTouch power supply" at either end. The 20-amp feeds are now run from the power source to the module feed area.

LiteTouch provides proper surge suppression units for its systems. These units vary in size and the mounting requirements are generally different for each separate unit. This surge suppression unit needs to be mounted and connected to the main service panel that services the

LiteTouch system (the same panel from which you ran the three 20-amp feeds). The closer this unit is to the service panel, the better. It is probably easier to install the suppression unit during the rough-in stage, but may be accomplished in most cases at some later time. Typically, it hardwires into a two-pole breaker inside the service panel. Each unit will have special instructions that will guide you through a proper installation. Lightning, surges, and spikes damage electronic equipment, as we all know. Anything that will decrease this risk will be helpful. There are surge suppression companies that will work with you if you need more protection for your system. You can never have too much. LiteTouch equipment incorporates transient surge suppression technology throughout its equipment. The transorb board (communication termination board) that is located in the control module enclosure is an example.

The electrical grounding system is an important part of a home automation installation. A separate grounding wire must be installed from a good grounding source to the CCU enclosure. This will later be connected directly to the CCU. Installing copper grounding rods is also recommended, as this reduces the overall resistance of the electrical grounding system.

The house lighting circuits are next (refer to Figure 1-12). Starting with circuit 101, we will progress through the home. Circuit 101 is an outside wall-mounted light located at the front door. A 12-2 NM cable is run from the electrical light box (circuit 101) to the right side of the control module enclosure inside the utility room. Mark the outer sheathing of the 12-2 NM with a 101 inside the control module enclosure. Sometimes the control modules are moved around inside the enclosure, so be sure to leave an ample amount of cable inside, just in case there are any changes. Bedroom #1 is where circuit 102 is located. Run a 12-2 NM cable from the outlet box to the LiteTouch control module enclosure in the same manner as you did in circuit 101. Mark both ends of the cable run with the proper circuit number. Run the rest of the 12-2 NM cables from the fixture and outlet boxes to the module enclosure in the same manner as were circuits 101 and 102, marking them correctly as you go. Mark the 12-2 NM cables inside the half-switched outlet boxes so they are not mixed up with the permanently powered cables for the other half of the outlet. You can choose to strip off the outer sheathing of the cables inside the control module enclosure; however, make sure you relabel the black wires with the proper circuit numbers. Cut a piece of cardboard to the size of the control module enclosure opening and place it inside. This will protect the interior of the enclosure from drywall powder, paint, and dust. It would be a good idea to double-check all of the work that was completed to make sure everything looks good. Check if the cabling is secure and protected against damage. Check that all the boxes and enclosures are secure and level. This concludes the rough-in phase.

# 2

# Finishing the LiteTouch 2000 Hardware Installation

## *CCU Install*

### Low-Voltage Terminations

Now that the drywall is completed and the interior of the house is painted, the rest of the hardware can be installed. The first place we will begin is at the CCU location. In the LiteTouch hardware package that was sent to you, you will find the Standard CCU (Figure 2-1 shows the rear panel of the CCU). Carefully remove it from its package and inspect the back of the unit. You will notice a row of four green connectors labeled CMP1 through CMP4. Remove one green connector and set it aside (it does not matter which port you take it from). You will also notice a row of orange connectors labeled CSP1 through CSP4. Again, remove one of the connectors and set it aside. Next, you will notice a single connector labeled Phone. Also remove this connector. Place the CCU back into its package so that it remains protected while you accomplish the low-voltage terminations to these connectors.

Locate the control module communication cable. It will be the only unmarked one in the CCU enclosure. It is also the only cable with the shielding foil visible at the cut end. Strip about an inch and a half of outer sheathing off the cable and separate the individual conductors. Take the green communication connector that you set aside earlier and position it so that the setscrew connections are facing toward you. Starting from the left-hand side of the connector, strip and attach the red

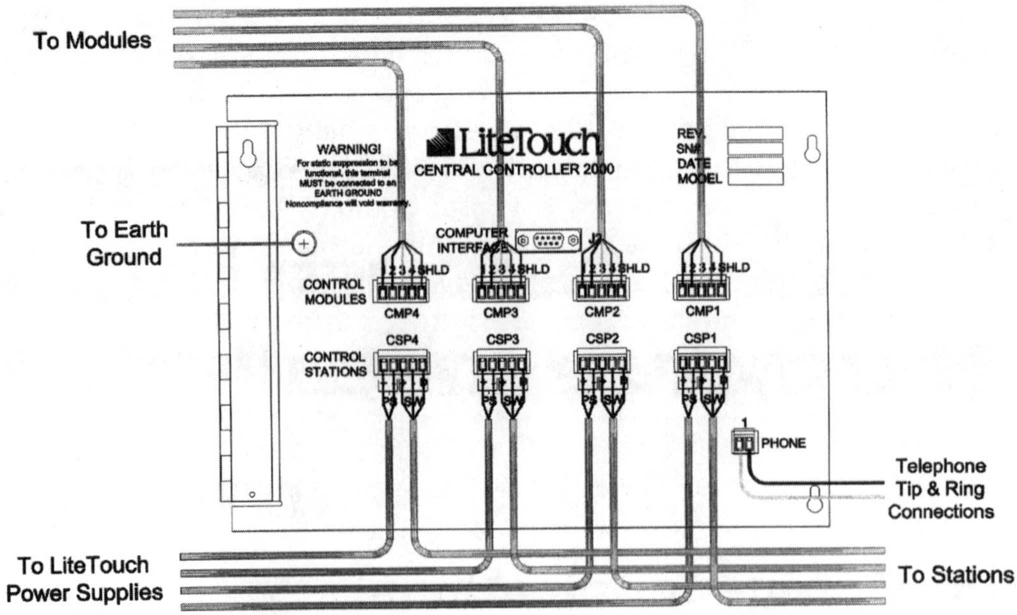


*Figure 2-1*

wire from the cable first. Next, attach the black wire, followed by the white wire, then the green wire, and finally the bare ground wire. Make sure that the connector terminals are snug and move the completed cable aside.

Locate the four-conductor unshielded cable that is marked, "From power supply." Strip about an inch and a half of outer sheathing from this cable. Retrieve the larger orange connector (the smaller one is the telephone connector) and position it so that the setscrew connectors are facing you. Using the red wire from this cable, attach it to the farthest left connection. Attach the black wire next, then tape or tie-wrap the blue and white wires back so they are not in the way. The blue and white wires will not be used in this cable. In the next step, you will need to find about a foot of four-conductor 16-gauge unshielded cable. Strip the outer sheathing of the cable completely and separate the red, black, and blue wires. You will use these pigtail wires next. Find the two switch station cables (one to the Data Input module and the other to the switch control stations) and strip away some of the outer sheathing like you did to the other cables. Strip the ends of the red, black, and blue wires from both cables and also the three pigtail wires that you prepared earlier. Wrap all the reds together (three wires) and wire-nut them securely together. Repeat this step for the other two colors. When you are finished, you will have three individual pigtail wires connected to the switch station cables. Prepare these pigtail wires again by stripping the insulation off the ends (just about an eighth of an inch). Take the orange connector and position it so that the setscrew connectors are facing toward you. In the third connection from the left, attach the red wire, then the black wire, and finally the blue. Tighten the connections securely. This connector is now complete.

In a larger switch station installation, you would run more control station cables from the CCU to the house switch control station boxes. Then you would evenly distribute the switch load among the other ports. More dc power will also be needed in this situation, so extra cables would be run from the power supplies in the module enclosure to the CCU. As a rule, it would be a good idea to distribute the same amount of power and switch control stations on each port. For instance, if you had 60 switches and four single power supplies, you would install four switch station cable runs from the CCU with 15 switches on each run. Then you would place each home run of 15 switches on each port with its own separate power supply on each port. This installation has only eight switches and one Data Input module to connect, so one port will suffice. The CCU only draws about 200mA, so this power requirement does not demand much from the dc power supply. The CCU is not a great consideration when planning the overall dc load.

The telephone connection is next. Retrieve the telephone cable and the orange CCU telephone connector. Strip the outer sheathing of the telephone cable back to about an inch and a half. If it is a two-pair cable, separate the green and red conductors, and strip and attach them to the orange connector. If the cable is a three-pair, match one pair of like-colored wires (one solid orange and one striped orange, for example). Attach them to the telephone connector, then go out to the telephone interface and mark the cable to let the telephone hookup company know what pair you want attached to the LiteTouch phone line. This labeling will save headaches later when trying to troubleshoot a nonworking CCU telephone interface problem. In many instances, if this labeling is not present, it will not get hooked up inside the interface box or sometimes the wrong conductors manage to get terminated.

Remove the CCU from its box and hold it up to the enclosure with the connectors facing the opening. Install the grounding wire to the supplied screw on the back of the CCU. Next, plug the orange control station connector into the CCU. Look carefully at the wires and the corresponding symbols below them. Make sure that the blue wire is above the letter "D" and that the next black wire from the switch station is above the negative symbol. Keep going across the connector and check if the red wire of the control station is above the positive symbol. The last two connections are the power supply and they should correspond to the black wire over the negative symbol and the red wire over the positive symbol. This is a good way to check if you wired the connectors correctly before you installed them into the CCU. Plug in the phone connector next.

The last connector that you need to install on the CCU is the green communication connector. Once you have plugged it into any one of the communication ports, do a check of the wiring by referencing the numbers on the CCU. The symbol, SHLD, should be above the bare ground wire. The number 4 should be above the green wire. The number 3 should be above the white wire. The number 2 should be above the black wire and the number 1 should be above the red wire. It should be noted that some equivalent communication cables have different colored wires in them (sometimes two or more of the conductors are colored black). In cases like this, it may be advisable to ohm-check and identify them before you start connecting them to the CCU. The colors of the communication wires used in this installation are from LiteTouch-supplied cables.

Tuck the cables and wires carefully back into the CCU enclosure and mount the CCU to the enclosure. We are finished with the CCU for now. The next area of installation will be the control module enclosure.

## Control Module Enclosure

Remove the cardboard protector and locate the communication cable from the CCU on the low-voltage side of the enclosure. Strip the outer sheathing back about 2 inches and separate the wires. Looking at the transorb board in Figure 2-2, which is located in the upper left-hand corner of the module enclosure, you will see the color codes for the wire attachments. Match the colors of the cable wires to the colors indicated on the transorb board. Secure the connections with a slotted-head screwdriver.

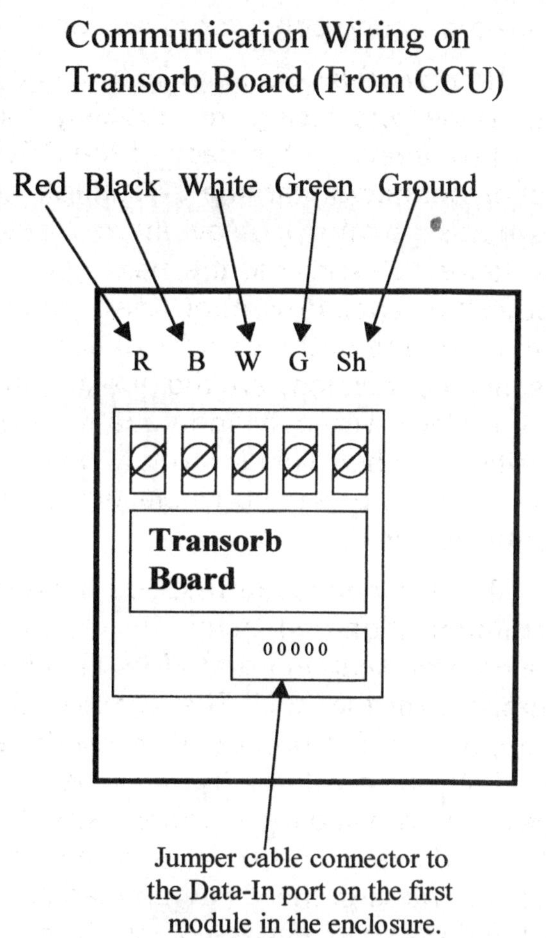

*Figure 2-2*

In the hardware package that was sent to you from LiteTouch, there are four module boxes. Open the relay module first. Insert the relay module into the top space provided and align it in the enclosure panel. Secure it to the panel with the two provided screws. Remove the dimmer module from its box and install it immediately below the relay module. Install the power supply below the dimmer and the Data Input module below the power supply. This is a good time to address the modules, so we will begin at the relay. From the enclosure schedule we see that the relay module address is 01, so this will be dialed into the MS (most significant digit) and the LS (least significant digit). Inspecting the left side of the relay module (Figure 2-3), we find the top address dial, which is the MS (this is marked on the module). Using a small slotted-head screwdriver, dial the MS to 0. Then, go to the lower dial (the LS) and dial it to 1. The relay module is now addressed. The dimmer module (Figure 2-4. Ignore the manual remote info for now.) will have an address of 02. Dial the MS to 0 and dial the LS to 2. The next module is a power supply

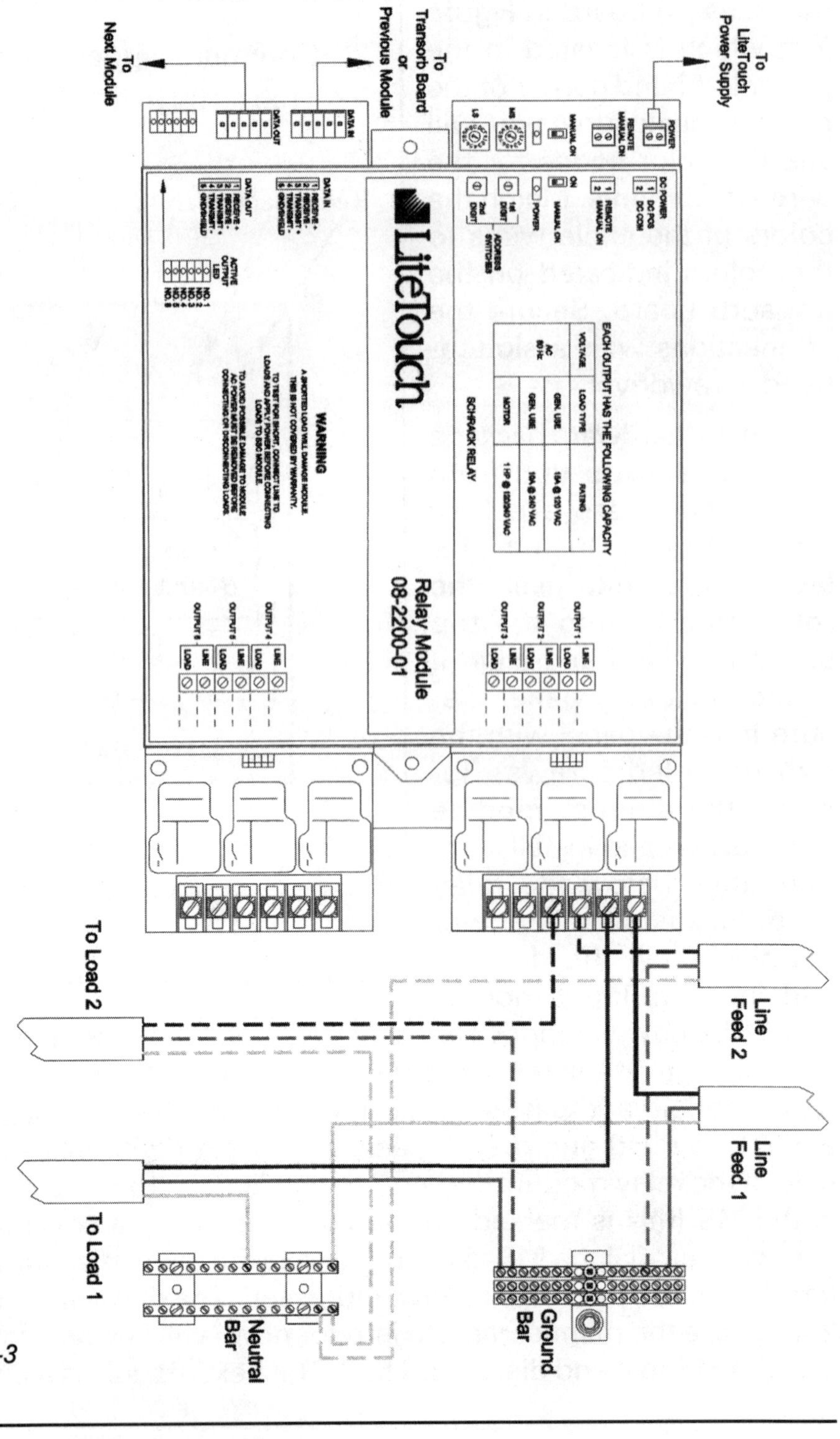
To LiteTouch Power Supply
To Transorb Board or Previous Module
To Next Module
LiteTouch
Relay Module
08-2200-01
SCHRACK RELAY
WARNING
Line Feed 2
Line Feed 1
To Load 2
To Load 1
Ground Bar
Neutral Bar

*Figure 2-3*

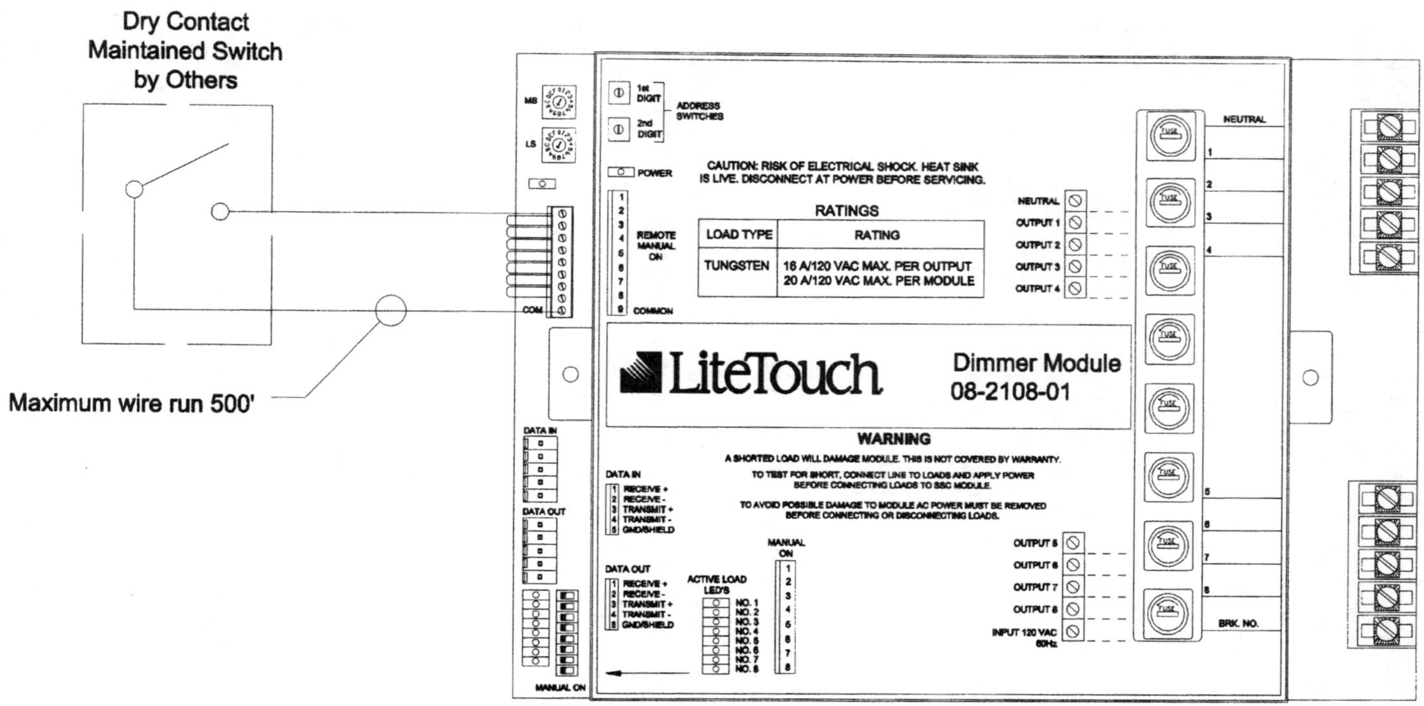


*Figure 2-4*

<u>Note:</u> Remote Manual On is shown with all inputs wired to one switch. Individual control may be used by wiring each input to individual switches. All switching is between the COM terminal and the Remote Manual On input terminals.

(Figure 2-5) and it does not have an address. The data input module (Figure 2-6) has an address, but it is not considered a true control module. This unit is, in reality, a switch. We will address it now, but we will use special addressing characters so it will be easy to spot among the regular switch control stations. At the MS (left dial), dial F and at the LS (right dial) dial F. The address for the Data Input module is FF and will be found as a switch by the CCU. It would be a good idea to label the module FF so it can be seen without reading the address from the MS and LS dials. The next step will be the low-voltage terminations to the left side of the installed modules.

The control modules should have been sent with communication jumpers attached to them at the Data-In port of each (see the Data-In and Data-Out ports back on Figure 2-4). Take the free end of the relay module jumper and carefully plug it into the transorb board. It is very easy to install these incorrectly, so we will go over a few points to avoid this. Make sure the wire colors on the jumper match the colors denoted on the transorb board. The red wire on the jumper connector should be closest to you when installed correctly. The bare ground wire on the connector should be near the back surface of the enclosure panel. In addition, it is easy to make a mistake and install the connector so that it is one pin off. Check to see if the connector hides all of the pins as you plug it in. Now, go to the dimmer module and take the free end of the communication jumper and plug it into the Data-Out port of the relay module. This concludes the communications between the CCU and the control modules.

The power supply low-voltage connections on this installation are fairly easy to do. Locate the cable that is marked "Power to CCU." Strip about 4 inches of outer sheathing from this cable and tape or tie-wrap the blue and white wires back. Strip the insulation off the ends of the red and black wires. Attach the red wire to the dc positive connector and attach the black wire to the dc negative connector. Tighten securely. In addition, locate and strip about 3 feet of 16-gauge cable and remove the red and black wires. Attach the red wire to one of the positive terminals of the power supply and the black to one of the negative terminals. Run the other ends of these wires up to the relay module. Attach the red wire from the power supply to the red wire of the jumper that comes out of the relay module. Then, attach the black wire to the black jumper

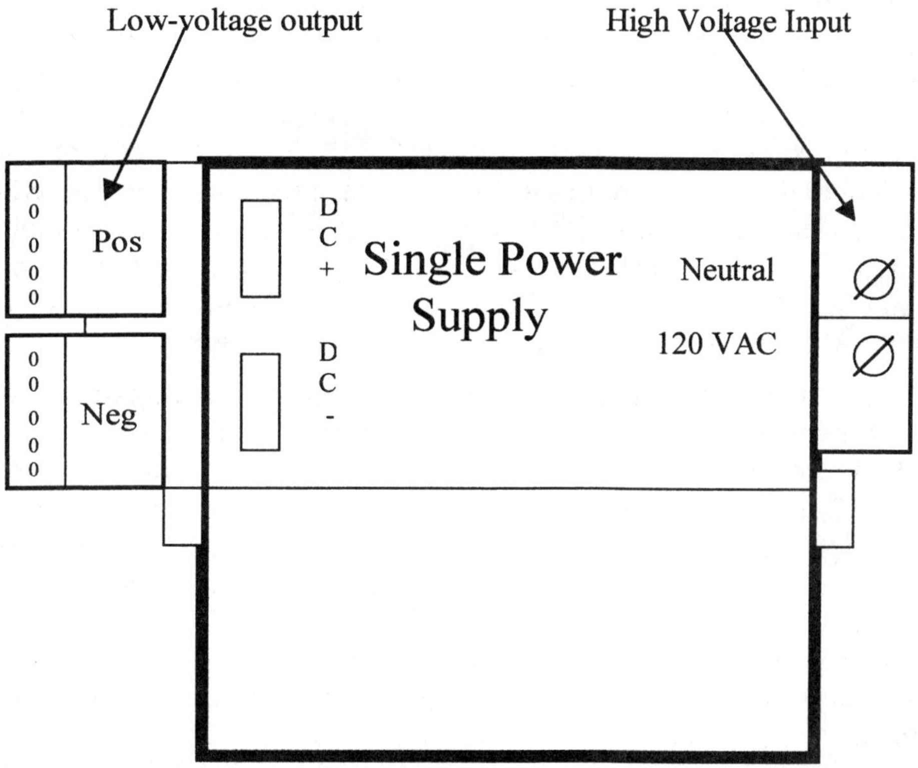


*Figure 2-5*

at the relay module. Use small wire nuts to secure and insulate the connection. The newer eight-channel relay modules do not require this dc power supply to operate. They are self-powered by the high-voltage input on the high-voltage side of the module.

In larger installations, there will be more power supplies and subsequently more cables run to them from the CCU. Dual power supplies are more common when more switches are implemented in a system. These dual dc power supplies can be connected in many different ways. Paralleling the dc outputs at the power supply is an accepted procedure for delivering more amperage to the system. This can be accomplished by installing jumpers between like polarities of different power supply modules and outputting anywhere along the parallel path. It should be noted that in some installations, the power supply is needed to power the dc side of certain control modules. In dc motor control relay modules, low-voltage relay modules and elec-

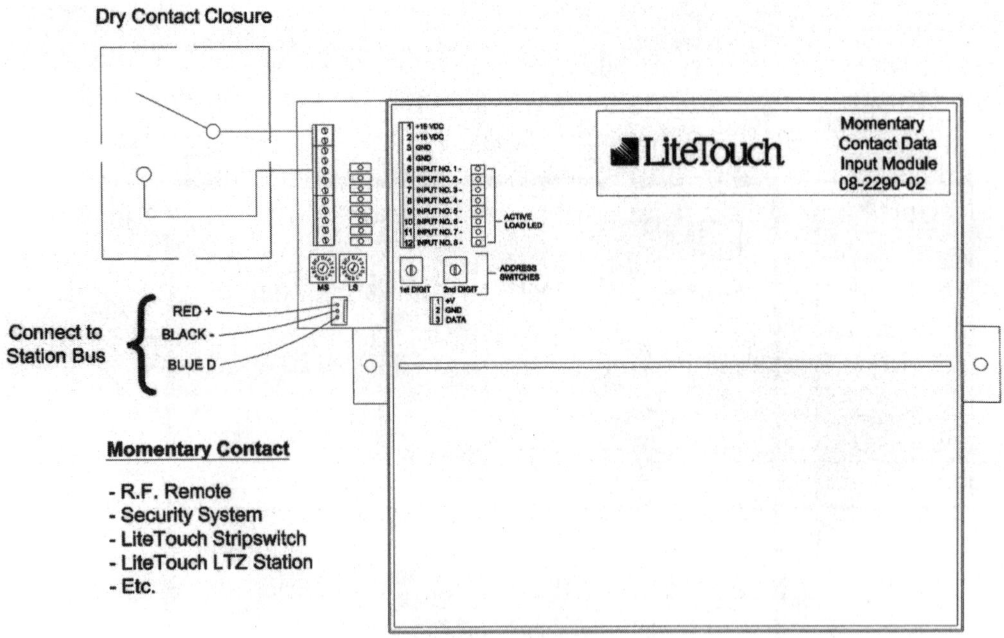


*Figure 2-6*

tronic fluorescent control modules, dc power needs to be supplied to them from the power supply module. Check the control module's specifications to make sure they are either self-powered by a high-voltage input or are dependent on a separate low-voltage input. In this installation, the dc power is now completed to the CCU and relay module.

The Data Input module will need to be connected to the switch station cable coming from the CCU. Locate this cable and strip about 4 inches off the outer sheathing. Wrap the white wire around the cable to keep it out of the way. This wire will not be used. Strip the insulation off the ends of the remaining wires. Connect the red wire of the cable to the red wire of the jumper connector that is attached to the module (see Figure 2-6 for the jumper). Repeat for the other two wires, matching color for color. Wire-nut the connections. The last low-voltage connection to be made is the security interface. Locate the cable that comes from the security box to the Data Input module area. Strip away a few inches of outer sheathing and connect the red wire to the number 1 terminal on the upper left part of the Data Input module (Figure 2-6).

Then connect the black wire to the number 5 position farther down the terminal block. If the security people are making the connections on the other end of this cable, make sure they use the red and black wires to attach to the single-pole dry contact activation relay. The address of this module is FF and the button number would be considered 1 by the CCU in this particular connection configuration of the Data Input. This would program as FF-1 in the CCU. This information will be needed later when it is time to program the CCU. This completes the low-voltage connections inside the control module enclosure. The high-voltage side of the control module enclosure will be discussed next.

## High-Voltage Connections

The high-voltage connections are located on the right side of the module enclosure partition. This installation procedure considers that all power to the panels is off. We will be using the same enclosure schedule as before, but it will be renamed as Figure 2-7 to make it a more convenient reference. At this time, it would be a good idea to mark the locations and load numbers on the modules. Self-adhesive labels work well for this. Start with the relay module at the top.

### *Relay Module 01*

Mark output 1: Outside wall light, #101.

Mark output 2: Bed. #1 receptacle, #102.

Mark output 3: Kitchen light, #103.

Mark output 4: Bed #2 receptacle, #104.

Mark output 5: Den receptacle, #105.

Mark output 6: Master Bed receptacle, #106.

Mark on front label of relay: Module 01.

Continue on to the dimmer module below it.

### *Dimmer Module 02*

Mark output 1: Foyer light, #201.

Mark output 2: Dining light, #202.

Mark output 3: Bathroom light, #203.

Mark output 4: Hallway light, #204.

## Control Module Enclosure Schedule

Module address: 01 Module type: Relay 6-channel No.of enclosures: 1

| MODULE POSITION | LOAD # | LOAD LOCATION | WATTAGE | CIRCUIT SUPPLY |
|---|---|---|---|---|
| 1 | 101 | Outside Wall Light | 100 | 1- 20 amp 110v circuit |
| 2 | 102 | Bed #1 Outlet | 200 | |
| 3 | 103 | Kitchen Light | 200 | |
| 4 | 104 | Bed #2 Outlet | 200 | |
| 5 | 105 | Den Outlet | 200 | |
| 6 | 106 | Master Bed Outlet | 200 | |
| | | Total Load: | 1100 | |

Module address: 02 Module type: Dimmer module

| POSITION | LOAD# | LOAD LOCATION | WATTAGE | CIRCUIT SUPPLY |
|---|---|---|---|---|
| 1 | 201 | Foyer Light | 200 | 1- 20 amp 110v circuit |
| 2 | 202 | Dining Light | 200 | |
| 3 | 203 | Bathroom Light | 200 | |
| 4 | 204 | Hallway Light | 100 | |
| | | Total Load: | 700 | |

| | |
|---|---|
| POSITION 3<br><br>POWER SUPPLY MODULE (SINGLE)<br>OUTPUT: dc | 1-20 AMP<br>110V CIRCUIT |
| POSITION 4<br><br>DATA INPUT MODULE (MOMENTARY)<br>CONTROL: Security Interface<br>Switch Station Address: FF-1 | N/A |

*Figure 2-7*

The next step will be to separate all of the 12-2 NM high-voltage cables into lines (from the power panel) and loads (from the fixtures and outlets). Pull the lines aside and strip the outer sheathings off of them. Be sure to mark what modules the lines will be feeding on the black wires once each cable is fully stripped. Neatly install the grounding wires to the grounding block at the top right of the enclosure. Then, neatly install the neutral (white) wires to the neutral blocks provided. Taking the black feed wire that is labeled "relay feed," attach it to output 1's line connection (in an 8-channel relay you would also run a neutral jumper from the neutral bar to the relay module). Take some of the excess black #12 wire and cut four, 4-inch lengths. Next, cut one more length of #12 wire to about 7 inches. Strip all of the ends of the #12 wires and bend them into a U shape. These will be your line jumpers. Attach one small jumper from output 1's line terminal and tighten the screw (the original feed), and one end of the jumper will be connected to this terminal. Place the other end of the jumper into output 2's line terminal. Attach another small jumper from output 2's line terminal to output 3's line terminal. Tighten output 2's line terminal screw. Attach the long jumper from output 3's line terminal to output 4's line terminal. Tighten output 3's line terminal. Continue this procedure with the jumpers until all six line outputs have jumpers to them. When this is completed, the feed connections will be complete to relay module 01. Strip the outer sheathing off the cable of load number 101. Attach the ground wire and neutral to the bars provided. Attach the black wire of load 101 to output 1's load terminal and then tighten the screw. Strip the outer sheathing off of the cable for load 102. Attach the ground wire and neutral to the bars provided. Attach the black wire of load 102 to output 2's load terminal and then tighten the screw. Continue this procedure with loads 103, 104, 105, and 106. When this is complete, relay module 01's high-voltage connections will be complete. Let's move on to the high-voltage connections on the dimmer module.

Locate and strip the 12-2 NM cable that feeds the dimmer module. Attach the grounding and neutral wires to their respective bars. Attach the black feed wire to the terminal on the module that is labeled "Input 120VAC." Attach a white jumper wire from the neutral bar to the neutral terminal on the dimmer module. This procedure allows the entire module to have power to it. No other jumpers are needed. Next, locate and strip the 12-2 load 201. Attach the ground and neutral to their respective bars as before and then attach the black wire (load 201) to output 1. Tighten the screw. Locate and strip the 12-2 load 202. Attach the ground and neutral as before and then attach the black (load 202) to output 2 on the dimmer module. Repeat these steps for loads 203 and 204 respectively. When all of these high-voltage connections have been terminated, dimmer module 02 will be fully complete.

The last remaining 12-2 NM feed to this panel is for the power supply. Strip the 12-2 NM feed and attach the ground and neutral to their respective bars. Attach a jumper from the neutral bar to the neutral terminal on the power supply. Attach the black feed to the 120 VAC terminal on the power supply. This will complete the entire control module enclosure for this installation.

There has been some concern as to the arrangement of the neutrals in this kind of installation. Some claim that all of the neutral feeds and loads should not go to one location. It mostly has to do with an effect known as parallel neutrals. It is important to realize that this procedure has been used for a very long time. Additionally, the enclosure and its components are UL listed, so the electrical techniques used need to be credible to receive this listing. The next section will be installing the LiteTouch control station switches.

## Installing the Control Station

In this section of the LiteTouch installation, we will reference the blueprint (Figure 1-12) for the placement of the house switch stations. Here is a list of control station locations and their respective ID addresses:

Foyer ................ Address 01 (5-button switch); Reference Figure: 1-3
Bedroom #1 .... Address 02 (3-button switch); Reference Figure: 1-4
Kitchen ............. Address 03 (4-button switch); Reference Figure: 1-5
Bedroom #2 .... Address 04 (3-button switch); Reference Figure: 1-6
Dining Room ... Address 05 (3-button switch); Reference Figure: 1-7
Bathroom ......... Address 06 (1-button switch); Reference Figure 1-8
Den .................. Address 07 (2-button switch); Reference Figure 1-9
Master BR ........ Address 08 (6-button switch); Reference Figure 1-10

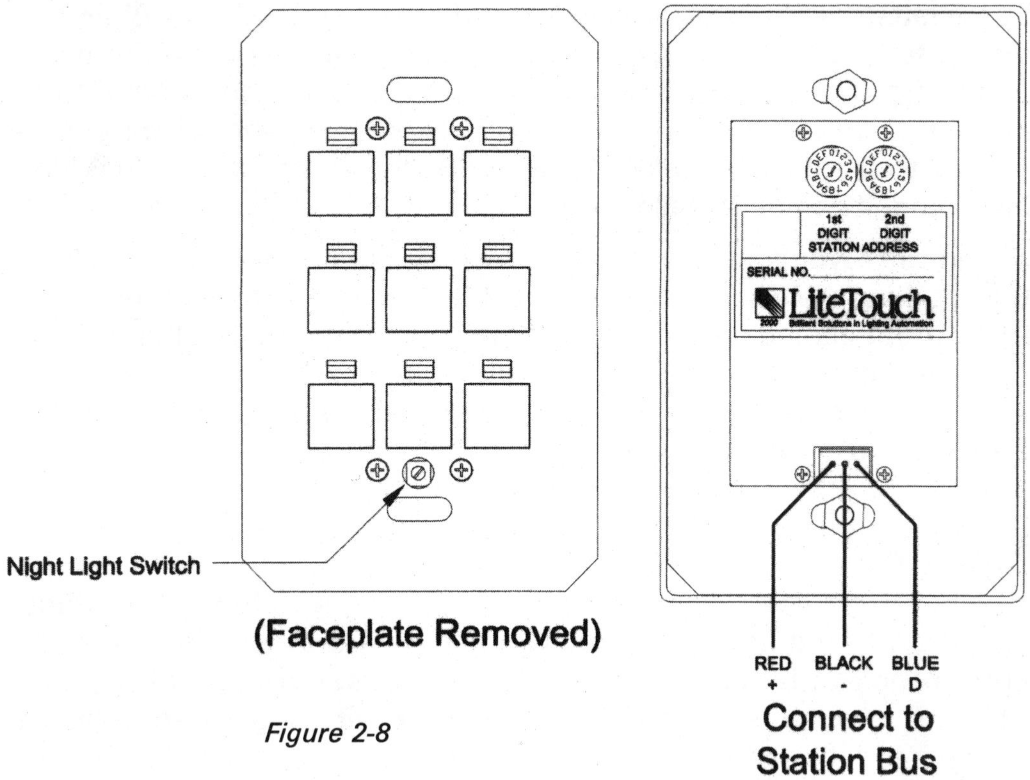


*Figure 2-8*

To stay organized, it may be a good idea to start at address 01 and progressively install the switches using the station addresses as a guide. In the hardware package that was sent to you, you will find the switch control stations. The outsides of the switch packing boxes are usually marked by their address numbers and sometimes by their specific locations. Generally, they are also addressed by their MS (most significant digit) and their LS (least significant digit). However, the switch stations will need to be double-checked to make sure they are properly addressed.

Open the packing box marked switch 01 and check the MS and LS address dials. The MS is usually indicated by the "1st digit" placard just below the dial (Figure 2-8). In some switches it is indicated by the letters "MS". The LS is usually to the right of the MS and is indicated by the "2nd digit" placard. As with the MS, the LS can sometimes be indicated by the letters "LS". Using a small screwdriver, dial the MS to "0" and dial the LS to "1". Switch station 01 is now addressed. Check the button engraving schedule on the switch station guide and compare it to the actual switch. The button positions and engravings should correspond exactly to the switch station guide. Once the switch station has been inspected, it can be installed.

In the foyer, locate the switch box that will house the switch station. Pull out the two switch station cables and strip off the outer sheathing of both cables (you can leave about an inch of sheathing on the cable as it enters the box). Separate the white wire from the red, black and blue. This white wire will not be used. Take the red wires, strip the ends off, and wire-nut them to the red wire of the supplied pigtail jumper (three red wires will be connected together). The jumper connector (shown in Figure 2-8) is usually taped to the outside of the switch packing box. Repeat this termination step for the black and blue wires. This type of connection is known as parallel. It continues the switch station circuit on to the next box and also allows the switch station of the current box to be connected to the switch station bus of the CCU. The red wire is the positive and the black wire is the negative. The blue wire is the data wire that communicates with the CCU. In some instances, the switch box may have more than two cables running to it. The connections are still made in parallel, so group all of the red wires and wire-nut them together and do the same to the blacks and blues, respectively. After these connections are made, plug the connector into the station. This connector is polarized by means of little notches on the side, so it will slide into the switch station connector only when properly aligned. Next, remove the magnetized cover from the front of the station and place it where it will not get damaged. Then, ease the connecting wires into the box while guiding the switch station into the box opening. Align the switch mounting holes with the box mounting holes. Secure with two 6-32 x 1-inch screws. Use a level to make sure that the switch is straight and then tighten securely. Check to see if the address numbers 01 are marked on the place provided on the switch. If these numbers are missing, use a pen and write those numbers on the switch. Next, place the magnetized front cover back onto the switch station and check that the plate lies flush against the wall. If it does not, you may have to remove the plate again and adjust the switch station until the cover rides flush against the surface of the wall. When this is finished, switch station 01 is completely installed.

Switch station address 02 will be installed in bedroom #1. Repeat the steps that were taken for station address 01 in the foyer. Double-checking the switch station addresses is one of the more important points when installing these types of switches. If one has been addressed incorrectly or placed in the wrong area, it will not work right. For instance, say that switch station 05 of the dining room was accidentally installed in bedroom #1's switch box. As you enter bedroom #1 and press the top button to turn on the floor lamp, you would instead be turning on the dining room light. If enough of these switches get swapped around, the house control would be pretty chaotic.

Repeat the previously explained installation steps for the remaining switch control stations 03 through 08. When the procedures for each switch have been finished, the hardware installation will be complete.

# 3

# Programming the Standard 2000 CCU

## Programming with a Personal Computer

Over the years since this system was first developed, different versions of programming software have emerged. Though these versions have appeared to be quite different from each other, they have had one common goal in mind: to make it easier to program switch and load information into the CCU. This task can be arduous depending on the size and complexity of the system being installed. The system design we have been working on up to this point is small and simplified, but it does demonstrate the sound principles that the system was based on. In this section, the steps that need to be taken to program the CCU from a personal computer will be explained. After the programming is complete, the actions required to upload this program into the CCU will also be demonstrated. When this version of programming has been covered, we will then program from the front keypad of the Standard 2000 system CCU. In the next chapter, another version of programming software will be addressed that is called LiteWare. LiteWare is a Windows-based graphical version that implements a design tool software with an additional program task. It can be used to program and export switch/load information for the Standard 2000 CCU. However, we will use it to program the 5000LC CCU. The 5000LC is an advanced load controller that can use most of the same system components that the Standard 2000 system utilizes. The 5000LC will be covered in the next chapter.

The programming for the earlier CCU software versions (many of which are still in use) was comparatively quite simple. After the computer drive prompt you typed C:> LT, and the program would execute and show you a couple of different items from which to choose. The number one item was CCU communications. This program would allow you to interface your PC with the standard CCU via the computer's RS232 ports. Uploading and downloading were accomplished in this way. The second choice was inputting switch and load data. This choice allowed the programmer to place the control for the addresses of both the switches and control modules into a format that the CCU could recognize. The newer version of this software will be covered next. When you learn this newer version, you will most likely be able to backtrack and use the older version if you found it necessary. It integrates some of the same commands and has similar programming inputs.

To obtain this programming software, you can visit LiteTouch's Web site at www.LiteTouch.com and download the version that you need. You can obtain a LiteTouch Electronic Resource Kit CD that also has this software on it. The specific installation instructions for your computer are given with each version requested. The programming steps discussed next will assume that the software has already been downloaded and installed on the personal computer. It may be helpful to have this software running on your PC while you proceed through this section. The programming steps will be more apparent that way.

Like many computer applications, there are numerous ways to travel through the environment to get what you want. The following example is one way; however, others may find a different way that suits them better. From your Windows environment, choose to use the MS DOS prompt. When this is accomplished, you may have to type **cd..** after the prompt to leave the directory of Windows. You will want to see the drive prompt all by itself ( for example: C:\>_). Type **cd\LT\PROG** after the drive prompt and then press Enter. It will then give you something that looks like this: C:\LT\PROG\>. Type **LTPMENU** after this prompt and press enter. You should then be inside the program looking at the main menu. If this command line does not work for you, try typing **LT** at the DOS C prompt (the LiteTouch CD install of this software uses this execute). If none of these commands works, check the LiteTouch directories on your drive to see if they are different from the ones mentioned here. You may need to change some of the directory information on your command line to arrive at the LiteTouch main menu.

At the main menu, there are many options listed, but for now we will choose the number 1 selection (standard programming). Once this is selected, it will ask you to enter the file name you wish to access. Type a file name that you would like to use (such as the owner's name). It will tell you that it could not open it (if it is a new file name) and would you like to continue? Answer yes. The program will then create this file. Another menu will now appear that lists five options. We will choose option number 1 once more (Switch Definition Menu). This option will allow you to modify the switch and load information. The switch definition menu will give you three items to choose from. From this menu choose the number 1 option again (View/modify switch DEF). You will then be able to program the switch and load information collected so far concerning the current installation in this section. Please reference Figures 1-3 through 1-10 for the upcoming practical programming section.

The field that will appear after you have entered "View/modify switch DEF", will say:

Sta Name___________ (Type **Foyer** on this line. Press Enter.)

Sta Address_________ (Then type **01** for the switch address)

Press Enter. This will drop you down to a nine-button switch figure. Press the arrow up key and scroll to the right until you arrive at the field that states the number of buttons for the switch. Type in **5.** A modified five-button switch figure will appear below and the number 1 will flash with a cursor. Press Enter from the flashing 1 and the cursor will move to the "Switch Function" field. Select Toggle from the list and then press Enter. When some types of new information are entered, the program will prompt and ask if you would like to create what you just entered because it could not find that information in the program. If the information you entered is correct, just press Enter and it will create and save it for you. Once you have pressed Enter from the Switch Function field, the cursor will travel across to the "Load Name" field. Type the words **Outside Light** and then press Enter. The cursor will then move to the "ADR" (address) field. Type in **01**. The cursor will then move to the "Output" field. Type **1** in this field. Engraving will be the next field. Looking back to Figure 1-3. You will notice that the engraving schedule states that the words "Outside Light" will be engraved on this button. In reference to this, type **Outside** on the first line and then arrow down and type **Light** on the second line. Press Enter to move the cursor to the "Save" position and press Enter again. This button is now programmed and the process will start over with the next button on switch 01.

The number 3 position on the switch will now have the cursor flashing on it. (Note that you are still programming switch station 01 in the foyer. You are only programming each of the buttons now.) Press Enter and the cursor will move to "Switch Function". On Figure 1-3, you will notice that this button is a dimmer, so select Dimmer from the list and then press Enter. The load name field will be next, so type the word **Foyer** and press Enter. The ADR field needs to have the first two digits of the load module that the switch button is controlling, so enter **02** in this field. This button controls the first load of dimmer module 02, so type **1** in the Output field. The engraving schedule of Figure 1-3 shows that the button will have the word "Foyer" written on it, so type this on the first line and arrow down to the second line. Nothing will be engraved on the second line, so press Enter to bring the cursor over to the Save field and then press Enter again. Switch 01-1 and 01-3 have been entered and saved.

This process moves along quite rapidly as you get used to the program. A couple of great things about programming on the PC in this way: it is fast and you can do it anywhere you have a computer. When the program is complete, you can move the PC (hopefully it is a notebook computer) to LiteTouch house and upload the finished program into the CCU, and that's all there is to it.

Switch 01-5 will be programmed next. After you have pressed Enter to save the information on button number 3, the next screen will now show the switch button arrangement with the number 5 button flashing. Press Enter and choose "Scene Preset" from the list. The Scene Preset program differs from the single load programs in a couple of different ways. These are the steps to program this type of control function. Press Enter and that will bring you to the Load Name area. Type **Kitchen Path**, then press Enter. The cursor will now be in the address area. Type 01. This will then bring you to the Output field. Type **3**. This will bring you to the "Level" field. Since load 01-3 is a relay module load, it can only operate at 100%, so enter **100%** and press Enter. The cursor will return to the Load Name field (you can type **Foyer**). Press Enter and the cursor will move to the address area again so that you may enter more load information. At this point, type **02** and then type **1** at the Output field. Now you are in the Level field and a decision has to be made as to the percentage amount that you will enter. This is a dimming type load, so amounts of less than 100% are expected. This load is the light in the foyer and this is a path light, so the percentage should not be set too low. Input **75** into this field and if the owner wants it changed later, it will not be difficult to do. Press Enter again to return to the Load Name field (you can type **Hallway**) and then press Enter. Now you are in the address field again. Type **02**, and then type **4**. Now we are back at the level field again. Considering that this load is the hallway light and also that this is a path button, 75% may also be a good dimming level for this situation. Type **75**, then press Enter. You are finished inputting loads into this field, but escaping this field is different than escaping the field of a single-load input. Press function key F3 and it will place the cursor in a field that is labeled Fade Time. Fade Time instructs the CCU on how fast it is supposed to bring a dimming load to its preset dimming level and also how fast it is supposed to turn the dimming load completely off. At this point, we will let the default settings remain and continue on to the engraving field. Press the arrow down key twice and this will bring you to the engraving field. Type **Kitchen** on the first line, then arrow down and type **Path** on the second line. Arrow right to go to the Save field. Press Enter and switch 01-5 is complete. When you are practiced at programming, you will be able to program a button like this in about five or six seconds. It is surprising how fast you can catch on to the routine.

Switch 01-7 is now highlighted with the cursor. Press Enter to bring it over to the Switch Function field. Scroll down the list. You will notice that there is an Off Master Toggle in the selection area, but no On Master Toggle. Figure 1-3 shows the switch type for 01-7 to be a Toggle-off and a Toggle-on type for 01-9. This station design guide was written partly with a software in mind named LiteWare (which will be covered later). In the LiteWare software, these toggle switches exist. However, this is not a problem in this software because we will program 01-7 to be a Master Off and 01-9 to be a Master On. The effect will be somewhat similar. Getting back to the programming, we were in the Switch Function list for switch 01-7. Scroll down and select Master Off from the list and press Enter. Type the load name as House Off and press Enter. Type **01**, then type **1**. You are now back in the load name field. The original load name remains the same, however, we have many more loads to enter on this button as shown in Figure 1-3. If a previous switch has these same loads already programmed into it, you can copy them to this area by use of the F9 key. Since this is the beginning of the program, and no such switch has all of these loads yet, we will place them manually. Press Enter to skip past the load name area and type **01**, then type **2**. Press Enter again and type **01, 3**. Press Enter, **01, 4**, Enter, **01, 5**, Enter, **01, 6**. Those are all of the entries for module 01. Now, we will program the four loads from Module 02. Press Enter, **02, 1**, Enter, **02, 2**, Enter **02, 3**, Enter, **02, 4**. Don't worry, this is much easier to program than it is to read. The other positive is that you will never have to enter all of these loads into a switch again in this program. Whenever you need to enter all the loads in the future, just press F9 and copy the loads from another switch into the current switch program. Now, to finish up switch 01-7, press the F3 function key to escape this field and enter the fade time field. Press the Arrow Down Key twice to arrive at the engraving field. Type **House** on the first line, arrow down and type **Off** on the second line. Arrow over to the Save field and press Enter.

The last button on this switch is 01-9 and it will be easier than 01-7 was. Press Enter to move the cursor into the Switch Function field. Scroll down the list until you come to Master On. Select it and move to the Load Name field. Type **House On** and press Enter. Instead of inputting these load addresses by hand, we will now press the F9 key. A box will appear that will say "From Control Station" and we will input 01 and then Enter. The next field wants the button number, so we type in our last button address that previously required all of the programming, 7. Now, cursor down to the "Copy" field and press Enter. All of the loads that were on switch station 01-7 are now copied to 01-9. F3 out of that field into the fade time field and double arrow down to arrive at the engraving field. Type **House** on the first line and **On** on the second line. Arrow over to the save and press Enter. Switch station 01 in the foyer is fully programmed on your personal computer.

When all of the data for the switches are programmed, it will then be transported to the CCU via the PC's RS232 port with an appropriate communications cable. For now, we will continue to program the next switch, which is control station ID address 02 (Figure 1-4).

Press the Escape key and this will bring you out of the programming for switch 01 and back you into the Switch Definition Menu. Press the number 1 key to program the next switch. You will again see:

Sta Name_____________(Type **Bedroom #1** on this line. Press Enter.)
Sta Address___________(Then, type **02** for the switch ID address.)

Press Enter. This will drop you down to a nine-button switch figure. Press the arrow up key and scroll to the right. (Depending on the situation, you may not have to scroll to the right. It may just arrive at the button number field when you arrow up.) Type in **3** at the button number field. A three-button switch configuration will appear below and the cursor will flash on the number 2 button. Press Enter from the flashing 2 and the cursor will travel to the Switch Function field. Scroll to Toggle and select it by pressing Enter. In the Load Name field type Room and then press Enter. The cursor will move to the ADR (address) field. Type **01** in this field and then type **2** in the Output field. The cursor will then automatically travel to the Engraving field. Type Room on the first line, arrow down, and leave the second line blank. Press Enter, or the arrow over key to arrive at the Save field and press Enter again to save this button program.

Button position 5 will have the cursor now. Press Enter to bring it over to the Switch Function field and select Toggle again. Press Enter and type the load name "Kitchen" into the field. At the ADR field, type **01** and then type **3**. The Engraving field will have only one line. Type **Kitchen** and arrow down and over to the Save field. Press Enter and another button is programmed on switch 02.

One more button program and control station 02 will be completed. It is important to keep the switch addresses and load addresses separate in your mind. It can be very easy to transpose the numbers and this will cause some strange activities when it comes time to test the completed system. Button number 8 will be flashing after the last button save. Press Enter and select Scene from the Switch Function field list. Press Enter and then type in **Bath Path** in the Load Name field. Press Enter and type **02** and then type **3**. This is a button that will be used mostly late at night, so a lot of light is not needed. Type **50** into the Level field and press Enter. This will bring you back to the Load Name field. Since the load addresses you are entering are all included in the Bath Path scene, press Enter to return to the ADR. Type **02** and then **4**. In the Level field, type **50** and then press Enter. Press the F3 key to escape to the Fade Time field. Leave the default setting by pressing the arrow down key twice, which will bring the cursor over to the Engraving field. Type **Bath** on the first line and Path on the second. Arrow over and save by pressing Enter. Switch station 02 programming is now complete.

Switch station 03 is located in the kitchen and the station design guide for it is illustrated in Figure 1-5. Press the Escape key to exit to the Switch Definition Menu and select the number 1 option again. You will see:

Sta Name______________(Type **Kitchen** and press Enter.)
Sta Address____________(Then, type **03** and then Enter.)

Arrow up and over to the Number of Buttons field and type **4**. Press Enter and the cursor will travel to the number 1 position on the switch figure. Press Enter and select Toggle from the Switch Function field. Press Enter and type **Kitchen** into the Load Name field. Press Enter and type **01** in the ADR and **3** in the Output. Type **Kitchen** in the Engraving field. Arrow down and over to the Save field and press Enter.

The number 3 button will flash; press Enter. Select toggle from the list and press Enter. Type **Bedroom #1** in the Load Name field and press Enter. Type **01** and then **2**. Type **Bedroom #1** in the Engraving field. Arrow down and over; press Enter to save.

Button 7 will flash; press Enter. Select Scene from the list and press Enter. Type **Kitchen Path** in the Load name field and press Enter. Type **01**, then **3**. Type **100** in the Levels field and press Enter (relays need 100%). Press Enter and type **02**, then **1**. Type **75** in the Levels field and press Enter. Press Enter and type **02**, then **4**. Type **75** in the

Levels field and press Enter. Press F3 to escape to the Fade Time field. Leave the programmed default by pressing the down arrow twice. Type **Kitchen Path** in the Engraving field (one word on each line) and arrow over to the Save field. Press Enter.

Button 9 will flash; press Enter. Select Scene from the list and press Enter. Type **Master Bed Path** in the Load Name field and press Enter. Type **01** and then **6**. Type **100** in the Levels field and press Enter. Press Enter and type **02** and then **4**. Type **75** in the Levels field and press Enter. Press F3 to escape to the Fade Time field. Leave the default setting by pressing the arrow down key twice. Type **Master** on the first line of the Engraving field and **Bed Path** on the second line. Arrow over and save the button by pressing Enter. Switch station address 03 is completed. The programming pace is getting quicker and your skills are increasing.

Switch station 04 is illustrated in Figure 1-6. Press the escape key to exit to the Switch Definition Menu. Press 1 and type **Bedroom #2** in the Sta Name field. Press Enter and type **04** in the Sta Address field. Press Enter and then arrow up and over to the Number of Buttons field. Type **3** and press Enter. Press Enter from the flashing 2 and select Toggle from the Switch Function list. Press Enter and type **Bedroom #2** in the Load Name field. Press Enter and type **01**, then **4**. Type **Room** in the Engraving field, arrow down and over. Press Enter to save.

Press Enter from the flashing 5 and select Toggle from the list. Press Enter and type **Kitchen**. Type **01** and then **3**. Type **Kitchen** in the engraving field, arrow down and over. Press Enter to save.

Press Enter from the flashing 8 and select Scene from the list. Press Enter and type **Bath Path**. Press Enter again. Press the F9 key (this is for copying a switch). In the box that appears, type **02** in the area that says From Control station. Press Enter and type **8** (this button will be a duplicate of 02-8). Arrow down to the Copy field and press Enter. Double-check to see if loads 02-3 and 02-4 are inserted into the address and output fields. F3 out of this area and leave the default by pressing the arrow down key twice. Type **Bath** on the first line of the Engraving field and **Path** on the second. Arrow over and press Enter to save. Switch station 04 is now programmed.

Switches 05, 06, 07, and 08 will be completed in the same manner as the ones that were just programmed. The last switch station that needs special attention is FF-1 and it is the Data Input module security activation address. The programming for this station will be a little different than the preceding ones. We will program input 1 to flash all the loads in the house on and off. This may intimidate an intruder. In the Sta Name field, type **Data Input** and then press Enter. In the Sta Address field, type FF and then press Enter. Arrow up and over to the Number of Buttons field and type **8** (the Data Input module has eight inputs) and then press Enter. From the flashing 1, press Enter. Scroll through the Switch Function list until you come to Timed Flash and then select it. Press Enter and then type **Security** into the Load Name field. Press the F9 key and the copy box will appear. Type **01** and press Enter. Then type **9**, then press Enter (switch 01-9 has all house loads programmed into it already). Scroll down to the Copy field and press Enter. Make sure all of the loads are present in the address and output fields. F3 into the Active Time field. This field will determine how long the lights will turn on and flash after the initial input from security occurs. Arrow over to the minutes field and type **15** (for 15 minutes) and press Enter. There will not be any engraving for the Data Input module, so arrow over to Save and press Enter. Timed Flash is a good function for this switch because security may have only one input to the Data Input module. The flashing sequence will time itself out and turn off, leaving it ready for the next security input. When switches 1 through 8 and the Data Input module are all programmed, we can continue to program more features into the system using the PC.

### *More PC Programming Features*

It is a great advantage to program from the PC rather than the LCD screen of the Standard CCU. The PC allows programming at an increased speed. In addition, it has capabilities that are not available in the standard CCU. One such capability is the programming of the astronomical clock. This clock can only be programmed from a PC at this point in time. Later CCU versions will probably implement this added feature. Once this clock information has been added into the PC, it can then be uploaded to the CCU. The main advantage of the astronomical clock is that it permits timing functions to be more accurate. This clock uses latitude and longitude statistics to dictate when certain loads will be turned on and off. With the exact geographical location of the LiteTouch house entered into the CCU, the guesswork for actual sunrise and sunset times will be eliminated. Here are the steps required to input the astronomical clock information into the PC.

The first program scenario is for CCUs that do not have the astronomical clock capability at all. The first step is to download the current program from the CCU into the PC. Uploading and downloading has not been covered yet in this section, but since this is a brand-new installation, there should not be any program information present in the CCU. Therefore, the program now residing in your PC will suffice as a downloaded file copy to work from. The next step is to upgrade the internal IC chips of the CCU to accommodate this change. For this step, you must call LiteTouch support at 1-800-527-5527 and relate to them your intentions. They will ask you the serial number of the CCU and what version software it has. They will then send you the instructions and the IC chips needed to accomplish your task. In most cases, when a new CCU is being used, these chips will already be installed, so there may not be a need to address this step in the astronomical clock installation. For the CCUs that do not have this capability, we will continue with this programming.

Starting at the main programs menu in the PC programming software, choose the number 7 selection (Standard CCU Astro Upgrade). The program will give you specific instructions as you proceed through. It will provide a helpful list to show what this program will do during the upgrade. This is the list:

1) Run the Definition Check utility on the DEF file to ensure correct format.
2) Convert the DEF file to a format compatible with astronomical time.
3) Create astronomical time information (be sure to write down this information for later use).
4) Incorporate the astronomical time information.
5) Upload the DEF file back into the CCU (see next section "Uploading to the CCU" for this info).

To accomplish the above sequence, you must have both of these two items:

1) The site location (city and state)
2) The address of one switch control station and button number

The program will then start and the first item it will address is the Definition Check. It will prompt you to enter the file name you wish to use. Type the one you have been using to program the switch stations. It will then complete the Definition Check by packing the memory and checking the offsets. Once this test has passed, it will then tell you to press any key to continue. After pressing any key, the program will announce that in the next step, the program will be converted to the format that allows astronomical time information to be incorporated. Press any key to continue. Once any key has been pressed, sorting and file saving information will be displayed. It will then prompt you again: Press any key to continue. The screen that comes up next will say:

Now the astronomical time information will be created. First, the latitude and longitude of your client's city needs to be located. Type city and press Enter:________(type the city name here and press Enter).

It will then prompt you to: Type the **2-letter state abbreviation** and press Enter:___(type the two-letter state abbreviation here and then press Enter).

The program will then display the coordinates of the city and state. Latitude:__________(for example: 37.59 may be displayed here). Longitude:_________(for example: 120.01 may be displayed here).

Below this line, the program will then prompt:

If the latitude and longitude are displayed above, press "C" to continue. If not, press "R" to retry or "A" to abort.

Press C to continue. The program will then prompt:

Enter the latitude listed above (1st number) in degrees and minutes:________(for example, you would type 37.59 here if that was what was displayed). Press Enter. The program will now prompt:

Enter the Longitude listed above (second number) in degrees and minutes:______(for example, you would type 120.01 here if that was what was displayed). Press Enter. The program will now prompt:

Enter the time zone: 5-Eastern, 6-Central, 7-Mountain, 8-Pacific. Select the number of the time zone where your client is located and press Enter. Once you have pressed Enter, it will then calculate the sunrise and sunset times for the file name that you entered previously by using the latitude/longitude information. The program will then tell you where this information is stored. The program will then give you the instruction:

Press "C" to continue if all information has been entered correctly or "R" for reenter, "A" for abort. Once you have pressed the C key, the program will give more instructions.

The astronomical time must be incorporated into the DEF file. The standard CCU interface program will be run. Complete these instructions:

1) From the Main Menu, press Enter 3-times.
2) Then enter an existing switch station (for example: 01).
3) Select button shape and number of buttons.
4) Type a number 1-9 to display a programmed switch. (this will be highlighted in the program).
5) To save the switch, hold down the Alt key and press the S key.
6) Then press the Escape key until the main menu is displayed again.
7) Then, exit to DOS by pressing 0.

This procedure will then continue when these steps are completed. The astronomical time will reside in the DEF file. Complete each of these steps, then the upload utility will do the final upload to the CCU.

At this point, the astronomical clock program has been saved so you do not have to do an upload right now. The program will lead you into the upload, but without the proper PC and CCU setup, it will not be able to establish communications. At this time, you should escape from this section. When you finally do complete the upload later on and you find that this program information is not included in the CCU, repeat these programming steps again.

Even though the CCU that you are working on may not need the preceding upgrade, you can initiate this program just to retrieve the client latitude/longitude information. This latitude/longitude information can then be used in the next section that only deals with the astronomical clock data entry. In many new installations, the astronomical data for your client has already been programmed into the CCU by LiteTouch. If this data has already been programmed into the CCU, these astronomical programming steps do not need to be completed.

The next program assumes that the CCU already has the astronomical clock capability and therefore only the clock information will need to be entered.

At the main programs menu, select number 6-Astronomical Clock Set-Up. This option will make sure that the latitude and longitude are entered into the program. After the selection has been made, the program will prompt:

1) Enter Latitude:________(enter the latitude that was calculated from the last program, then press Enter)
2) Enter Longitude:________(enter the longitude that was calculated from the last program, then press Enter)
3) Enter Time Zone: 5-Eastern, 6-Central, 7-Mountain, 8-Pacific. Pick the number of the time zone you are in and press Enter. The program will then prompt:
4) Enter the filename that you want this information entered in:________(type the same file name you have been using to program the switches). Then press Enter. The program will now tell you that the information that you have just entered will be stored in (your filename.ast ) file. You can now press escape to return to the main menu.

The next option on the main programs menu is 8-Convert DEF/CDF to SWP. This program converts either a .DEF or a .CDF file extension to a .SWP. When you complete this short program, it will make the necessary conversions to the filename that you input.

Option 9-Standard Definition Check will be covered later on in the text. You may remember that the Definition Check has already been used during the astronomical upgrade.

Option A-All CCU Modems, allows you to communicate with the LiteTouch CCU at one of your customer locations. The menu items will allow you to either manually dial the phone number or choose a number from a stored list. In addition, it gives you the option to inform

LTMODEM to which communications port your modem is connected. The All CCU Modems option can only be used with LiteTouch CCUs that are version 1.70 or later. Additionally, the CCU you are calling must have modem capabilities.

Programming the timing functions will be addressed next. From the main programs menu, select 1-Standard Programming. It will then prompt you to enter a file name. Once the proper file name has been entered, the main menu will then appear. Select 2-Timing Menu from this list and press Enter. Various options will appear, select 1-Timer Settings. Timing fields will then be displayed that will look something like this:

Date Type:

( ) Days

( ) Calendar date

Time Type:

( ) Clock Time

( ) Sunrise

( ) Sunset

Selecting the first field (Days), will display a second field to the right that will allow you to input each day of the week. All days can be entered by pressing the F2 key, permitting a drop-down list to appear. Each named day of the week can be scrolled with the arrow keys and then selected with the Enter key. When all desired days have been selected, press Escape and then press Enter to arrive at the Time Type field. If you selected Calendar date to begin with (either Days or Calendar date can be selected, not both), you would have received a field to the right that let you select any named month and any numbered day of that month. Pressing Enter from this field area would also bring you to the Time Type field. At the Time Type field, you can select one of the options. Choosing the Clock Time would make the load turn on or off strictly using the standard time. Selecting the Sunrise or the Sunset option would implement the astronomical clock feature (either of these two selections will allow you to set a value at XX minutes either before or after the sun event). Move the cursor to the field that you desire and press Enter.

The next screen will appear and it will prompt you to select either an On control or an Off control. Use the arrow key to select one, and press Enter. Another field will then appear to the right that prompts you to choose a Switch Station. Type in the **two-number digits** of the switch you would like to control (for example, switch 01 in the Foyer). Then, type in the **one-digit button number** (for example, type 1 for the outside light). This timing control would turn on or off (whichever one you selected) control station 01-1. It will activate this selection for whatever day or date you chose. In addition, it will activate 01-1 at any clock setting you set for it (for example, at 6 p.m., or at sunrise or sunset). As you can see, this gives you many timing options to chose from for switch station 01-1.

After all of the information has been entered for this single control, press Enter and the previous screen will return. The cursor will then rest on the Save tab. Press Enter to save this timing information. You can then continue to input more timing controls into these fields. If you programmed an On for the last switch station, you may want to program an Off for the same switch station. This program is excellent for timing many types of controls such as HVAC, pool heaters, walkway lights, shutters, etc.

From the Timing Menu, select 2-Vacation Mode and the Times and Loads selection screen will appear. From this screen you can create a "Lived In" effect while the client is away from home. Load start and stop times are listed, as well as the loads (enclosure address loads) that they will control. Input the hours and minutes along with an A.M. or P.M. designation into the Start and Stop fields. Arrow to the right to input the desired load names or addresses. Press F2 for a list to choose from. Every time you enter a value into one of these fields, the cursor will automatically jump to the next field, so this will save some inputting time. Press Enter on the Save tab when you are through with the vacation mode programming. Press Escape to return to the Timing Menu.

Scroll down to the third selection, 3-Daylight Savings and then press Enter. You can either choose daylight saving or standard time. Select whatever time you are currently in, then press Enter on the Save tab when complete. Escape back to the Timing Menu.

Select 4-Astronomical Clock from the Timing Menu and press Enter. This screen permits you to inspect the latitude, longitude, and time zone settings. Compare the numbers and selections with the astronomical information you saved from the earlier clock programming to verify that what is displayed on the screen is correct. Select Save and then press Enter.

This concludes the explanation of the menu items on the Timing Menu. The next PC programming items that will be covered are located in the File Menu section.

From the main programs menu, select 1-Standard Programming. The main menu will appear, select 3-File Menu from the list and press Enter. The File Menu will have six selections. From these selections, choose 1-Read, Create DEF File. This option reads a file stored on disk. A directory and file name need to be entered to retrieve the desired file. Escape to return to the File menu.

Selecting 2-Save Current DEF AS, will allow you to save the current file as another name. In effect, this will make a duplicate copy of the current file. Press Escape to exit to the File Menu.

Selecting 3-Back up DEF to Floppy will supply you with the drive number for the floppy so you can back up the current file to disk.

Selecting 4-Restore DEF from Floppy will retrieve a file from the disk in your floppy drive.

Selecting 5-Print Reports will print information regarding switch programming, load assignments, timer programming, and systems parameters. To the right of these selections, it gives two choices as to the way you would like to print either the switch or the load information. It either lets you sort the order by the switch or load name or by the switch or load address. This just means that switches and loads will either appear numerically in order (by address) or by name in order on the printed page. Once you have made your selections, press Enter on the Print command tab. These printed reports can be very handy when you are walking through the house checking switch stations and enclosure loads. It can also be a great help if you are programming the 2000 CCU from the front keypad instead of the PC. CCU programming from the front keypad will be covered later in this programming section.

Selecting 6-Printer Setup, will permit you to select which type of printer you are using and from what port that printer works. Press Enter on the Save tab to affect changes.

The last section that will be covered in PC programming will be the System Settings Menu. Access this menu by starting at the main programs menu. Select 1-Standard programming and the main menu will then be displayed. Select 4-System Settings Menu. This Menu has six different options as follows:

| | |
|---|---|
| 1. Computer Interface | 4. Switch Options |
| 2. Telephone Interface | 6. Preset Lock |
| 3. Interlock Type | 7. LED Invert |

The number 5 is not used in this menu. Selecting 1-Computer Interface permits you to set the parameters for interfacing the personal computer to the CCU. From the interface screen, press Enter at 9600 baud to select it. The cursor will move to the next field automatically, which is Parity. Select None and press Enter. The next field is Stop Bits. Select 1 and then press Enter. The cursor will now move to the Protocol field. This field gives two choices, CRC and ASCII. You may want to check with LiteTouch support to make sure for your particular installation, but the default is usually ASCII, so that is what we will choose here. Go to the Save tab and press Enter. Escape back to the System Settings Menu.

Selecting 2-Telephone Interface will allow you to choose the parameters used in the modem/telephone system. The first option on the Telephone Interface screen is the Security Active field. If this is marked with an "X", it means that the system is currently secure from unauthorized entry. Removing this "X" will turn security off. Ordinarily this will be an owner preference, but most will want it activated. Scroll down and enter a 10-digit security code. This code will be requested whenever the LiteTouch telephone system has been dialed up. Make sure this code is given to the owners, they may even want to change it themselves later on. The next field is the Answer on Ring Number field. Depending if this system has a dedicated telephone line or not will often dictate how many rings are entered into this field. One ring is fine for a dedicated line. More rings will generally be needed if another device is the priority telephone line user. Usually the owner can help you with this decision. Enter the amount of rings and then scroll down to the next field, which is Answer Mode. Answer Mode will give the option of Voice or Modem. Voice Mode allows the on/off switching of loads over the telephone line. Modem Mode allows a remote computer to send/receive the system definition. In addition, Modem Mode also has voice capability. Make the selection that is appropriate, then go to the Save tab and press Enter. Escape back to the System Settings Menu. The next selection on this menu concerns the interlock function.

Option 3-Interlock Type, will permit you to either select an eight- or a six-channel relay for interlock use. Interlocking relays are commonly used on motor-type loads. For instance, a house may have motorized shutters that are controlled by LiteTouch. There is usually a three-wire high-voltage cable running to the shutters from the control module for motor control. One wire is the neutral and the other two wires are switched power for the motor's forward and reverse movement. If power ever came across both switched wires at one time, it would most likely burn out the shutter motor. An interlock relay mechanically isolates these two forward and reverse power leads so that when one lead is energized, the other lead is locked away from the power. The earlier interlock relay modules were exclusively six-channel types. LiteTouch continues to carry these six-channel interlock relay modules, but in most cases specifies the eight-channel relays for new installations. The eight-channel relay allows more loads to be controlled in one enclosure, so in the long run, it can save money to specify these. This option will allow you to specify what type of interlock relay is installed so that the CCU can be programmed accordingly. When you have specified what type, save and escape back to the System Settings Menu.

The next selection on this menu is number 4-Switch Options. Choosing this preference will permit you to set the dimmer options for the ramp time and the minimum turn-on level. In addition to these dimmer settings, it will let you change the Scene Toggle to an "On" only function. There is also another area that lets you change the Timed Toggle function to an "On" only function. An "X" in front of each selection will, in essence, be a Yes to what function is stated there. If you change the defaults in this section, you need to save them before you escape back to the System Settings Menu.

The menu skips the use of the number 5 and goes straight to 6-Preset Lock. This selection will permit you to lock or unlock scene preset levels. In the unlocked position, scene levels can be changed at the actual switch station locations. If it is locked, the levels can only be changed at the CCU or with a file uploaded from a PC. Oftentimes, a scene preset is programmed with many separate dimmer loads. With an unlocked system, each of these loads can be set manually by individual dimmer buttons. Once all of the loads are dimmed to their desired levels, you press in and hold the scene preset button for four seconds and the new levels are then set in memory. This allows the homeowners to effect programming changes without actually knowing how to program. Once these changes are completed, it is a good idea to go to the CCU and lock the system. Many times when the system is left unlocked, someone will press and hold in a scene preset button while all of the dimmer loads assigned to that button are off. The scene preset button will now act like an Off Master. All the loads will now be set to the new level of 0% and will go to that level when the scene button is pressed. This is a common problem when a new system has just been installed. During the PC programming phase, it is a good practice to leave the system locked. It can be unlocked at the CCU later when this option may be needed. Escape back to the System Settings Menu.

Selection 7-LED Invert is the last item in this menu. It will allow you to change the status of the LEDs on the switch station buttons. Normally, the LED turns on when the load itself turns on. If you elect to invert, the LEDs will then react in an opposite manner. Many people opt for this choice for one reason or another, but it is often an owner preference decision. This will need to be discussed with them prior to changing this default programming. Pressing Enter at each function in this list will reverse the LED status. The list is as follows:

___Dimmer ___Raise ___Lower ___Timed Toggle ___Timed Flash

___Toggle ___Open/close ___Master On ___Master Off ___Scene Preset Toggle

___Momentary ___Dimmer Master ___Master Off Toggle ___Vacation

If you do not want to change the defaults, press Escape to return to the System Settings Menu.

Sometimes programming certain options later at the CCU may be a favorable choice. For example, the owners may have preferences that are not known to you when you are initially programming their house. When you set up a meeting with the owners after their house has been completed, you can then fine-tune the system to suit their special needs.

The next section of this chapter will cover the PC interface with the Standard 2000 CCU. Uploading all of the programming that has just been completed will be addressed first.

### *Uploading to the Standard 2000 CCU*

The personal computer that is used to do the actual program uploading to the CCU needs to have an RS232 port that has no other peripherals assigned to it. The next requirement is to acquire a proper communications cable to accomplish the upload from the PC to the CCU. LiteTouch has these cables available; however, you can modify a standard RS232 nine-pin cable if you desire. The cable and pinout in Figure 3-1 is used only for the Standard and Compact CCUs. The programming that has been accomplished to this point has been for the Standard CCU, so this cable will work fine. The cabling and pinout configuration in Figure 3-1 will be needed for this CCU installation.

Modifying a standard RS232 cable is a simple case of swapping pins 7 and 8 at one connector (as seen in Figure 3-1). You will notice that pins 1, 4, 6, and 9 are not used in this cable. Now that the preparations for the personal computer have been accomplished, it is time to go back to the house and prepare the CCU for the upload.

The house should have the power on when you arrive. The electricians should have checked their circuits and made sure that the lines and loads did not have any shorts in them. Make sure all of the circuits to the LiteTouch equipment are turned off before you begin.

Remove the CCU from its enclosure or swing it out so you are looking at the low-voltage connections on the back of the unit. Remove the green communications connector with the attached control module cable from the back of the CCU. Also, remove the orange power and switch station connector and cabling. Turn on the circuit breaker for the power

View is looking at connectors head on.

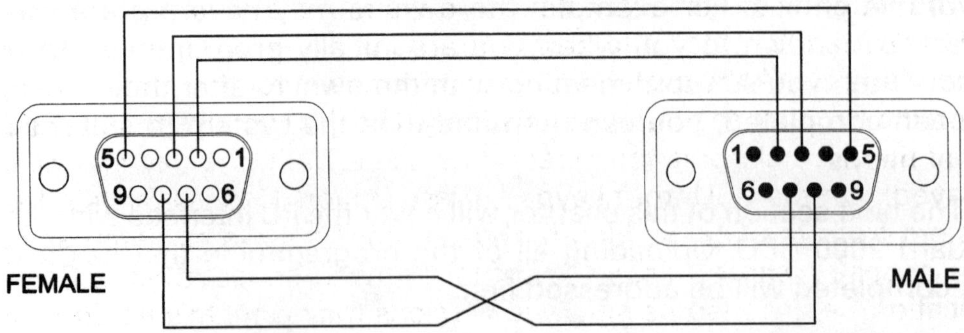


COMPUTER SIDE PINOUT

Pin 2 - Rec. Data
Pin 3 - Trans. Data
Pin 5 - Signal Ground
Pin 7 - RTS
Pin 8 - CTS

LITETOUCH PINOUT

Pin 2 - Trans. Data
Pin 3 - Rec. Data
Pin 5 - Signal Ground
Pin 7 - RTS
Pin 8 - CTS

Note: Pins 1, 4, 6, and 9 are not used by the Standard or Compact.

*Figure 3-1*

supply only and return to the CCU. Using the dc voltage setting on your multimeter, touch the positive probe to the setscrew of the red wire coming from the power supply and the negative probe to the setscrew of the black wire from the power supply. These wires are easy to identify in this installation because the power supply wires come directly from the cable that has the blue and white wires taped or tied back. The voltage measurement should indicate a dc voltage between 18 and 20 volts. The voltage may indicate a little higher value than this since there is currently no load from the control stations yet. If the voltage checks out, turn the circuit breaker to the power supply back off. At the CCU, plug in the orange power supply and switch station connector. The next step will be to set up your personal computer and CCU uploading cable. With your PC off, plug one end of the cable to the dedicated RS232 port and the other end to the RS232 port located on the back of the CCU. Turn on your computer and access the LiteTouch PC software as was explained earlier in this chapter. When you arrive at the main LiteTouch programs menu, select

9-Standard Definition Check. It will then prompt you to enter the file name that you want to check. Enter the same file name that you previously used to create the switch and load data. The program will then pack the memory and check the switch and load information for any errors. Next, turn on the circuit to the CCU power supply. The CCU should initialize and display the normal operation screen (the screen with the day and date displayed). On the CCU front keypad, press Option. Choose System Parameters from the Option Menu. Then, from the systems parameters menu, choose Computer. From the Computer menu, choose Change Communication Settings (some software versions may label this as Change Communications Protocols). The next screen will want you to select a baud rate. Choose 9600. The next screen will ask you to select the parity and stop bits. Choose no parity and one stop bit. After this setup is complete, the CCU should return automatically to the Options Menu. Now, press the Back key on the front of the CCU until the normal operation screen appears. The CCU is now ready for the upload.

From the main programs menu on your PC, select 4-Standard/Compact Communications. The program will ask what port you will be using. Enter the port number. Next, it will ask the baud rate you intend on using. Choose (1)-9600. The PC will then try to establish communications between itself and the CCU. Once communications has been established, it will then display a menu. Choose to Upload to the CCU from among these selections. It will then ask for a file name. Type in the same name you used for the definition check. When the correct file name has been entered, it will then start the upload. At this time, the display screen of the CCU will show that the computer interface is in use. When the upload is finished, you can then disconnect the RS232 cable and exit the LiteTouch program from your PC. The CCU is now programmed. Replace the control module communication connector and reattach the CCU to the enclosure. It may be a wise decision to also turn the power supply circuit breaker back off until you are ready to test the entire system. In this powered-down state, there is no need to worry about losing the file information you just uploaded. The CCU will retain its memory for an extensive period of time without power input.

It should be noted that downloading from the CCU is also possible. Once communications has been established between the PC and the CCU, a downloading option becomes available. At the end of the CCU download, the program will ask you to assign a name to the file. Once this file name is entered, the downloaded file resides in your PC. There are several reasons to have this downloading capability. If the current CCU residing program needs to be changed, it is much easier to effect these changes using the PC versus the CCU keypad. It is also much faster to program and pack the memory using a PC. In addition, running the definition check is always a better idea when the program resides in the PC. If you are at a LiteTouch installed house that you have never visited before, it is a good practice to download the house program onto your computer. In the future, that same house may be struck by lightning and this could damage the CCU or program file. In this situation, you will have a backup copy to reinstall.

### Checking the Programming

Now that the program has successfully been uploaded to the CCU, it may be a good time to see how the home will respond to the program. To do this, power must be turned on to the entire LiteTouch system. In addition, the individual enclosure components need to be checked. Before the three circuit breakers for this system are turned on, inspect the control module enclosure for anything out of place. Check for any loose connections or wires that are not attached to their proper equipment terminals. Double-check with the electrician that all of the loads and lines are free of shorts and that the high-voltage system is working properly. After this thorough examination has been completed, turn on the three breakers to the system. Go to the enclosure and inspect the control modules. The green power LEDs on the relay, dimmer, and power supply should be illuminated. The Data Input module does not have a green LED to show power status. The operation of this module will be examined later.

With your multimeter set to dc, measure the voltage on the left side of the power supply (low-voltage side). You should receive a reading between 15 and 20 volts. Fifteen volts is a rather low value, especially in a small system such as this, so anticipate a higher value. When this reading is complete, manually turn on the relay and dimmer modules. This is accomplished by turning on the override switches on the left side of the relay and dimmer. You will then see the red LED load status indicators illuminated on both modules. Now, check through the house to see if all loads are turned on. If some are not working, check to see if the fixtures have lamps installed in them. We will not cover troubleshooting in this section in any great detail. This will be covered in a later chapter, so we will assume that the problems will be minimal. Once the loads have been determined to be activating with the manual overrides turned on, go back to the enclosure and turn the overrides back off. Secure the enclosure door so that no one can access it. If other people are in the house, they may not know that the panel is live, so closing it is a good safety precaution.

The next step will be accomplished at the CCU. Remove or swing open the CCU from its enclosure so that you have access to the back termination area. With your multimeter set to dc, measure the power supply (PS +, –) voltage on the orange port connector. In addition, measure the positive and negative supply going out to the switches (SW +, –). Both of these measurements should have the same value (between 15 and 20 volts). You can reference back to Figure 2-1 to verify the control station port markings on the back of the standard CCU. After these measurements have been taken, place the negative probe on the negative port connector going to the switches (SW –) and the positive probe on the data port (SW Data). The voltage measurement should not be more than 1 volt less than the first (SW +, –) dc measurement. Typically, a healthy data voltage will read under 1 volt less. For example, if the first (SW +, –) measurement was 18 volts, the data measurement (SW –, D) should be around 17.2 to 17.5. If the measurement you received is close to this example, most likely you will have good communication from the switches to the CCU. With these measurements completed, reattach the CCU to the enclosure and secure. Some system tests on the CCU will be performed next.

### *System Tests*

Press the Option key on the CCU and then press the number 2 key (system tests or, on some versions, install tests). Three choices are then displayed:

1) Toggle Switch Display Flag (now off)
2) Control Module Communications Test
3) Control Station Communications Test

The number 1 selection is not used too often. To start this test, press the 1 key on the front of the CCU. To disable the test, press the 1 key again. While this test is enabled, you can go to a switch station and press one of the buttons. The LED status will then change (from whatever it was before you pressed it). Go back to the CCU and more information about that switch will be displayed on the screen. The switch station address, the port number that the switch is connected to, and the button position pressed will all be displayed on the screen. This test verifies if a particular switch is operating correctly in the system.

Selecting number 2 will test communications to the control modules. This test runs up to 99 loops of the control module system for response verification. If no problems are encountered, it will give a report at the end of the loop test. In this situation, you will receive a message like this:

Test Results (press any key to continue)

01 Reported Every Loop (this means the relay module at the top of the enclosure is working fine)

Press any key and there will be a report on the dimmer module in the second slot of the module enclosure.

Test Results (press any key to continue)

02 Reported Every Loop

In this installation, the control modules have passed the communications test. Press Enter and the screen will return to the Install Tests Menu (on some software versions this will be the System Tests Menu).

There is another testing tool for the control modules called a Power Bus Monitor. This testing tool checks the communications between the control modules and the CCU by using the data-out port on one of the control modules. The Power Bus Monitor tool also checks the integrity of the control module wiring. The standard CCU control module test may show that there are no problems with communications. This result may not necessarily be true because the CCU will always try to talk around a communications fault. This is actually a good design feature for normal operation, but not for a testing situation. Using the Power Bus Monitor will result in a more accurate control module communications test. More information about the Power Bus Monitor will be included in the chapter on troubleshooting.

The next test to be performed at the CCU will be for the switch stations. From the Systems Tests Menu (or Install Tests Menu), choose number 2 (System Tests). From this list, choose number 3 (Control Station Communications Test). This test includes five parts and can take a little while to complete.

1) Each control station port is checked for integrity.
2) The LEDs on all of the switches are turned off.
3) A communications response is requested from the stations.
4) All of the LEDs on the stations are turned back on.
5) A second response is requested from the control stations.

Make sure that no one presses a switch button while this test is in progress. It will be logged as an error and it will make the test results inaccurate. A warning is given at the beginning of the test that pertains to this. The screen will then prompt you to "Press any key to go on." When you press a key, a message will appear that says, "Pressing Cancel will abort station test." It will then start testing the CCU ports. If a problem arises, the display may warn you that a particular port is non-functional. Press any key to go on. In this particular installation, it is only important to pay attention to errors pertaining to the one port that was used. The other three empty ports will not have any effect on the used port, since all ports are isolated. We will discuss this test in more detail later on in the Programming the CCU from the Front Keypad section.

The next part of the test will turn off all of the LEDs to the control stations. After this is completed, the next part of the test will start a search for all of the control stations. During this test, two possible error messages can occur: Station Reporting Out of Order or Station Reporting for the Second Time. The Reporting Out of Order message may mean someone has pushed a switch button in the house during the test. It can also mean that the control station cable run is experiencing interference problems with the data line. The error message Station Reporting for a Second Time usually means that more than one switch has the same address dialed into its MS and LS digits.

The next part of this test will turn on all of the LEDs. After this has been completed, it will search for the control stations again. This search may come up with three error messages—Station Reporting for a Second Time, Station Reporting Out of Order, and Station Not Reporting with LEDs On. The first two error messages have already been explained. The third message often means that the LEDs have brought the power of the switch station line down low. Therefore, this indicates that there is not enough power in the switch station line to accommodate any extra load stress (such as every LED being on). This problem can sometimes be solved by adding a power supply to the switch station line (LiteTouch prefers that you consult it before taking this action).

When all of the control station tests are completed, a list of every switch station is displayed. Check through this list to make sure switches 1 through 8 are included. Also, make sure that the Data Input module FF is on this list.

As mentioned earlier, troubleshooting will be covered in more detail in a later chapter. At that time, specific test results and problem solving will be addressed more thoroughly. For this installation, we can assume that all of the tests have passed. Using the Back key on the CCU, press it until you arrive at the normal operation screen. The next step will be to go to the switch stations and check the actual operation of the system.

### Verifying the Switch and Load Functions

Now that the switch stations and control modules have been tested at the CCU, it is time to verify that the switches will actually activate the correct loads. Switch 01 is located in the foyer, so we will start there. Figure 1-3 shows that it is a five-button switch. The first button is labeled "Outside Light." Press this switch and make sure that the LED has illuminated. Look out the front door and check that the light has also turned on. Press the button again and verify that the LED has now turned off. Look outside again to make sure that the light has turned off. Proceed to the button labeled "Foyer" and press it. The foyer light and LED should activate. Press the button again and the light should turn off along with the corresponding LED. Now, press and hold the same button and the foyer light should ramp (get brighter or dimmer). Let it cycle through the entire dimming range and then let go of the button when the desired level is reached. Tap the button and it will turn the light off. Tap it again and the foyer light will illuminate to the same dimming level that you just set. If it works within these parameters, the programming is correct. The next button on switch 01 is labeled "Kitchen Path." Press this button and the light in the foyer, hallway, and kitchen should turn on. The foyer light will probably dim to a different level than it did for the second button. This is normal because the CCU instructions tell it to do this. Press the button again to turn off the lighting for this path.

Each time you test a button, be sure to check the LED load status indicator on the switch. It should correspond to the same state as the load. The next button will be labeled "House Off." Skip this button right now and we will get back to it after the next button on the switch is tested. Go to the button labeled "House On." Press it and every load in the house should turn on. Take your time and check every outlet and fixture in the house to make sure this switch is, in fact, turning every LiteTouch load on. Return to the switch in the foyer and press the button labeled "House Off". Check the house again to make sure every load is now off. If everything checks out, switch 01 in the foyer has been programmed properly and is communicating with the CCU.

Switch station 02 is located in Bedroom #1. It should have three engraved buttons and the first will be labeled "Room." Press it and check that the half of the outlet that is controlled by LiteTouch is operating (one half of the outlet is wired hot all of the time). Press the button again and check that one half of the outlet has turned off. Proceed to the next button labeled "Kitchen." Press it and make sure both the kitchen light and the LED indicator are lit. Press it again to turn it off and go to the button labeled "Bath Path." Press it and check that the hallway light and the bathroom light have turned on to a fairly low level (it was programmed to 50%). Press the button off. If all is well, proceed to switch station 03 in the kitchen.

The Kitchen switch station is a four-button type and the first button will be labeled "Kitchen". Press it and check for load and LED activation. Press it off. Go to the button labeled "Bed #1." Press it and check the bedroom #1 outlet like you did earlier. Press it off. The next button is labeled "Kitchen Path." This button was created to turn the lights off once you arrived from the foyer. This switch will also work well for the reciprocal when you walk from the kitchen to the foyer. In this instance, you can then turn the same loads off at the foyer switch. Check to make sure that the kitchen, hallway, and foyer lights turn on and off from this location. The next button is labeled "Master Bed Path." Press it and check that the hallway and half of the outlet in the master bedroom has turned on. Press it off. This concludes the test for switch 03.

There may be a situation where a button press does not make the intended load turn on. In cases like this, the best way to determine the problem is to check with the CCU to determine what may be wrong. From the normal operation screen, press Enter. Press 1 and then press 1 again. Then press Enter. A message will display on the screen that states that the next button that is pressed four times will be identified. Go to the switch button that is not turning the load on and press it four times. The LED will pulse, showing the CCU has accepted the switch press. Go back to the CCU and it will display the switch station address and a couple of options below it. Make note of the switch address and check that it is correct. Select the first option: View and Modify Switch. It will then display what function the switch has and the load(s) that it controls. Make a note of the load address and check to make sure that it is correct. In a new installation, this will more than likely point to the problem. The assigned load or switch address may be wrong. Find the right address and input it at the current display screen.

Continuing with the switch programming checks, proceed to switch station 04 in Bedroom #2. The first button labeled "Room" should control the half-switched receptacle. Press it on and off while checking the outlet with a high-voltage tester. Go to the next button labeled "Kitchen" and press it on and off. The kitchen light should respond accordingly. The last button is labeled "Bath Path" and this should control the hallway and bathroom lights.

The dining room is next (station 05) and it is a three-button switch station. The first button is labeled "Dining Room" and it is programmed as a dimmer. Press and hold the button while it cycles through its entire dimming range. Stop pressing in at the desired setting and this release will set the level. Press the button on and off to verify the setting and LED status. The button labeled "Kitchen" is next. Press it on and off to see if the kitchen light responds. Go to the last button labeled "Hallway." This is a dimmer, so repeat the steps for checking the dimmer operation (the steps used on the first button of this switch).

The bathroom has a one-button switch (address 06) and it is a dimmer. Repeat the steps required to test the dimmer and load program.

The den has a two-button station (address 07). The first button is labeled "Den" and it controls the half-switched outlet in the room. The second button is programmed as a dimmer and it controls the hallway light. The labeling on this button is "Hallway." Test both buttons to see that they conform to the programming that is required.

The master bedroom is a six-button switch station and it uses address 08 as an ID. The first button is labeled "Room" and it controls the half-switched outlet in this room. The second button is labeled "Hallway" and is programmed as a dimmer. Test both of these switches in the same manner as before. The third button is labeled "Kitchen Path" and it is programmed as a Scene Preset. This button controls the path from the master bedroom to the kitchen. It controls both the hallway and kitchen lights. The fourth button is labeled "Bath" and it is a dimmer that controls the bathroom light.

Test the third and fourth buttons to see if they operate their loads and LEDs properly. The fifth button is labeled "House Off." Skip this switch and return to it after the last button on this station has been pressed. The sixth button is labeled "House On" and controls all of the loads in the house. Press it and check that every load in the house turns on. Next, go back to the House Off button and press it. Make sure all of the loads in the house respond by shutting off. This concludes the checking of the standard switch stations in the house. The only switch station left is the Data Input module (address FF).

It would be helpful if you could access the security box, so an appointment scheduled with the security people on the day you are at the house would be a good idea. Ask security to activate the dry contact relay that the Data Input wires are connected to inside their box. Upon activation of the dry contact relay, the house should flash all of the loads for 15 minutes and then go back to the original switch settings. This may disturb the other building trade people in the house, so warn them ahead of time.

This simplified home automation system is complete. It could use some additional fine-tuning and this will be illustrated as the book continues. This particular installation will be used as a reference for the programming of the Standard 2000 CCU from the front 20-point keypad. In addition, this installation information will also be used to program the 5000LC CCU from the LiteWare environment in the next chapter.

## Programming the CCU from the Front Keypad

Figure 3-2 illustrates the front display and keypad of the Standard LiteTouch 2000 CCU. Keys 0 through 9 are used for control station and module address inputs. In addition, the CCU will prompt when one of these keys is needed for some other purpose. Keys A through F are called control station alpha address keys. These keys have other uses, but generally not in the normal programming environment. The Option key will transfer you to the Options menu and the Cancel key will transfer you back to the normal operation screen. Finally, the Enter key is used throughout the programming environment for a variety of tasks.

The programming that we are about to do assumes that the CCU has not been programmed yet. Programming from the CCU keypad will be similar to the PC programming that was previously completed. The menu system, however, is structured quite differently. There are advantages to programming from the CCU rather than a PC. One is that you do not have to carry your PC with you to effect changes in the CCU. The astronomical clock programming is one of the few tasks that you cannot accomplish without a PC, so nearly all required programming can be completed by inputting the data at the 20-point keypad.

During regular CCU operation, the normal operation screen will be displayed. It will look very much like it is depicted on the LCD display in Figure 3-2. The clock and date will probably be wrong until you enter the correct one later. Press Enter from the normal operation screen—this screen will be displayed:

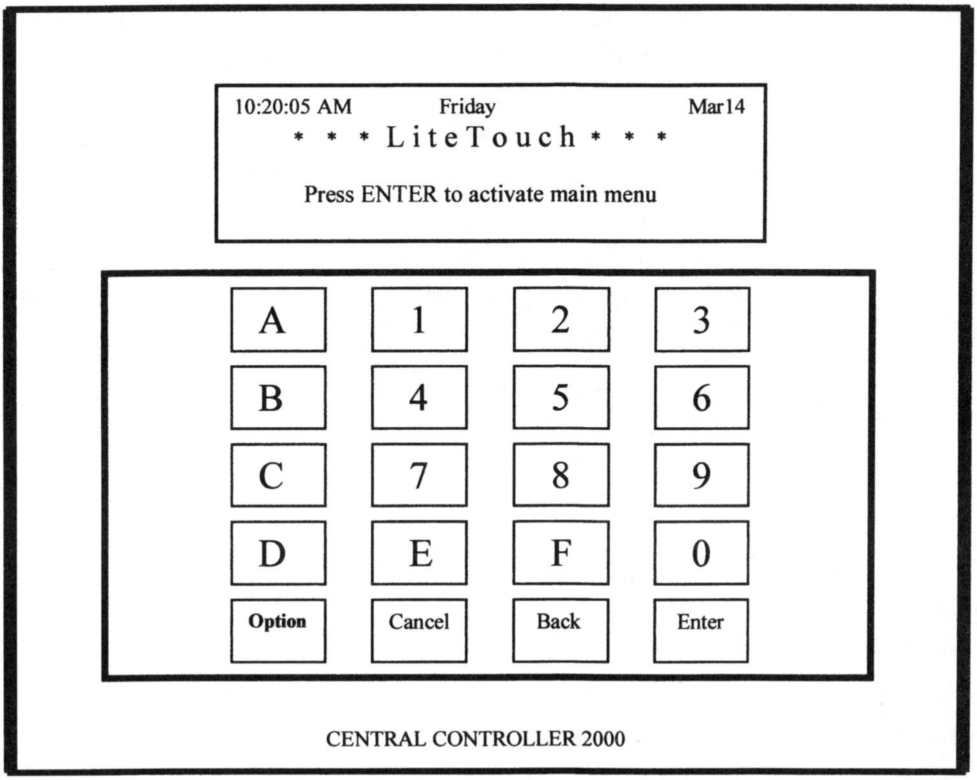


| | |
|---|---|
| 0-9; | Control Station and Control Module Address Keys |
| A-F; | Control Station Alpha Address Keys |
| Option; | Press to Enter Option Menu |
| Cancel; | Press Cancel Key in any Menu to get back to Main Menu |
| Back; | Press Back Key to go Back a Screen |
| Enter; | Press Enter Key to Enter Input into CCU. Use Only when Asked for on the screen |

*Figure 3-2*

MAIN MENU (Free Memory 98%)
1-Switch and Load Menu 3-Set Clock
2-Timer Function Menu
-Back space for Normal Operation-

The Free Memory value will change as you enter data. After the programming is complete, the memory will be packed and it should show a higher percentage value at that time. The Switch and Load Menu selection is where most of the programming will take place, so that is where we will begin. Please reference Figures 1-3 through 1-10 while we program the switch and load data into the CCU.

### ***Programming the Switches and Loads***

In this section, the switch and load data will be programmed into the CCU from the front keypad. These CCU switch/load inputs will be somewhat similar to the PC switch/load programming that was performed earlier in this section. CCU inputs will not progress at the speed of PC programming. Many technicians, however, prefer inputting at the CCU because of the easy-to-follow menus.

Press the 1 key to select the Switch and Load Menu. This menu will then be displayed:

SWITCH AND LOAD MENU
1-Examine/Update Switch Assignment
2-Examine Load Assignment
3-Unassigned Load Find

Press the number 1 key to select Examine/Update Switch Assignment. The CCU will then prompt you:

CONTROL STATION ADDRESS ->_
Input station address and press ENTER
or just press ENTER for Switch Identify.

There is no need to identify a switch yet. We will go over this option later. Input the first switch station address, which is 01 in the Foyer. Press 0 and then press 1. (You will find that on switch station and load addresses with a zero at the beginning, that they really do not need the zero entered, but adding it is a good habit to get into.) After pressing 0 and 1, press Enter. The next screen that will appear will look like this:

SWITCH POSITION NUMBER->_
Input switch position (1-9) and press
ENTER.

Since this (Foyer) is a five-button station, the first button position to program is 1. Press 1 and then press Enter. If this switch station address is already programmed into the CCU, it will arrive at a totally different screen once you have pressed the Enter key. For this situation, we will assume that the CCU is devoid of any switch and load data, so in regard to this, the screen that will appear next is:

SWITCH 01-1 is Unassigned
Do you wish to assign it?
1-Yes 2-No

Press 1 for Yes. The next screen will then appear:

SWITCH TYPE?

| | | |
|---|---|---|
| 1-Dimmer | 2-Open or Close | |
| 3-Raise/Lower | 4-Toggle | 5-Momentary |
| 6-Timed/Flash | 7-Master | 9-Scene Preset |

This switch button will activate the Outside light, and it is controlled by a relay module. So, press 4 for Toggle. The next screen will then display:

CONTROL MODULE ADDRESS->_
Input control module address and
Press ENTER

The load address for the outside light is 01, so press 0 and then press 1. Press Enter and this is the screen that will be displayed next:

OUTPUT NUMBER->_
Input the module output number and
Press ENTER

The load output number for the outside light is 1, so press 1 and then press Enter. The next screen will then prompt you:

Do you want to save these inputs?
(You can easily modify them later.)
1-Yes 2-No

Press 1 for yes. This screen will disappear and you will be transferred back to the Switch and Load Menu screen. This entire process must be accomplished for every button. The pace will quicken as you progress, but care needs to be taken as your input speed increases. It is easy to get ahead of yourself or start transposing load addresses for switch addresses. It can get kind of messy trying to find the entry errors afterward. The process starts over with this screen:

SWITCH AND LOAD MENU
1-Examine/Update Switch Assignment
2-Examine Load Assignment
3-Unassigned Load Find

Press the 1 key again from this menu to program the next button. The next screen will show:

CONTROL STATION ADDRESS->_
Input station address and press ENTER
or just press Enter for switch
Identify.

The next control station address will be 01, so press the 0 and 1 keys again. Press Enter and the next screen will appear:

SWITCH POSITION NUMBER->_
Input switch position (1-9) and press
ENTER.

The Foyer light is controlled by button number 3, so input this and then press Enter. The next screen will then display:

SWITCH 01-3 is Unassigned
Do you wish to assign it?
1-Yes 2-No

Press 1 for Yes. The screen will now display the function choices:

SWITCH TYPE?

| 1-Dimmer | 2-Open or Close | |
|---|---|---|
| 3-Raise/Lower | 4-Toggle | 5-Momentary |
| 6-Timed/Flash | 7-Master | 9-Scene Preset |

Press 1 to select the Dimmer function. The next screen will appear:

CONTROL MODULE ADDRESS->_
Input control module address and press ENTER.

In Figure 1-3, the Foyer light is assigned to the dimmer module 02. Press 02 and then press Enter. This screen will then be displayed:

OUTPUT NUMBER->_
Input the module output number and
Press ENTER.

On dimmer module 02, the Foyer light is connected to the first output. Press 1 and then press Enter. Then the last screen will display:

Do you want to save these inputs?

(You can easily modify them later.)

1-Yes 2-No

Press 1 for yes and you will be returned back to the Switch and Load Menu. If an inputting mistake is made before you save the switch button information, press the Back key. This will back you up a screen so you can start the input again. If a mistake has been made and saved, this next example will demonstrate how to change it. We will begin at the Switch and Load Menu.

SWITCH AND LOAD MENU

1-Examine/Update Switch Assignment

2-Examine Load Assignment

3-Unassigned Load Find

Select the first option again. This will bring the control station address prompt back.

CONTROL STATION ADDRESS->_

Input station address and press ENTER

or just press Enter for Switch Identify.

Input the switch station address where the error was entered. Let's say it was our last switch station entry of 01. Press 01 and Enter. The next prompt will be:

SWITCH POSITION NUMBER->_

Input switch position (1-9) and press

ENTER.

The last button position entry was 3, for the Foyer light, so press 3 and then press Enter. The next screen that will appear is one that has not been displayed yet. It is the screen for changing the information on an existing switch:

VIEW - MODIFY/DELETE SWITCH ASSIGNEMT

1-View and Modify Switch Assignment

2-Delete Switch Assignment

We won't do anything as drastic as deleting the switch, so select 1. The screen that will now be displayed will look like this:

SWITCH 01-3 (ENTER for no change)
1-Change Type (Now a Dimmer)
2-Change Assigned Load (Now 01-2)
3-Set Min/Max Fade Levels (Now 0/100%)

The mistake that was made is now obvious. The assigned load address was transposed. It is now 01-2 and it should be 02-1. Press 2 to change this error. The next display will be:

CONTROL MODULE ADDRESS->_
Input control module address and
Press ENTER.

Press 02 and then press Enter. The next display will be:

OUTPUT NUMBER->_
Input the module output number and
Press ENTER.

Press the 1 key and then press Enter. The change has now been saved. The next screen should be the View-Modify/Delete Switch Assignment display. Check to see if the load address is now correct, if it is, press the Back key until you are back at the Switch and Load Menu again. In regard to the switch station that was just edited, if there was a need, you could have easily changed the switch type or the fade levels while you were in that change environment.

Sometimes, you may feel you are messing up a program. If that happens, press the Cancel key. Additionally, if you are lost somewhere in the menus, press Cancel and this will return you to the normal operation screen (the one that gives the time and date).

The next button on switch station 01 is position 5. This will now be programmed into the CCU from the Switch and Load Menu:

SWITCH AND LOAD MENU
1-Examine/Update Switch Assignment
2-Examine Load Assignment
3-Unassigned Load Find

Press the 1 key. You will then be prompted:

CONTROL STATION ADDRESS->_

Input station address and press ENTER
or just press ENTER for Switch
Identify.

Press the 0 and 1 keys, and then press Enter. The next prompt will be:

SWITCH POSITION NUMBER->_

Input switch position (1-9) and press
ENTER.

Press the 5 key and then press Enter. The unassigned switch prompt will then appear:

SWITCH 01-5 is Unassigned
Do you wish to assign it?
1-Yes 2-No

Press the 1 key and the switch function choices will appear on the next screen:

SWITCH TYPE?

| | | |
|---|---|---|
| 1-Dimmer | 2-Open or Close | |
| 3-Raise/Lower | 4-Toggle | 5-Momentary |
| 6-Timed/Flash | 7-Master | 9-Scene Preset |

Press key 9, Scene Preset. This selection will require some extra programming screens in comparison to a single load/function button. The next screen will then appear:

CONTROL MODULE ADDRESS->_

Input control module address and press ENTER.

The kitchen path scene preset has three loads to enter. We will input the first one now, so press 01 and then press Enter. The next prompt will be:

OUTPUT NUMBER->_

Input the module output number and press ENTER.

Press the 3 key, and then press Enter. The next screen will display this:

Do you wish to assign
any more?
Press 1-Yes 2-No

Press the 1 key for Yes. The screen will then prompt:

CONTROL MODULE ADDRESS->_
Input control module address and press ENTER.

Press the 0 and 2 keys, then press Enter. The prompt will be:

OUTPUT NUMBER->_
Input the module number and
Press ENTER.

Press the 1 key. The prompt will then be:

Do you wish to assign
any more?
Press 1-Yes 2-No

Press the 1 key for yes. The screen will then display:

CONTROL MODULE ADDRESS->_
Input control module address and press ENTER.
Press 02 and then press Enter. The screen will prompt:
OUTPUT NUMBER->_
Input the module output number and press ENTER.
Press the 4 key and then press Enter.
Do you wish to assign
any more?
Press 1-Yes 2-No

Press 2 for No. All of the control module addresses and load numbers have now been entered. There is still some more programming to do with this scene preset button. The next screen will display this:

Fade Time (Minutes:Seconds) ->_
Input minutes (0 to 66) and press
ENTER. Then input seconds and press ENTER.

Any minute amount is a long fade time. I am sure you can imagine turning off a light and waiting one minute for it to slowly dim to off.

However, there are some practical uses for long fade times. Press 0 and then press Enter (for 0 minutes). Then press 02 and press Enter. For this installation, the fade time will be 2 seconds. The next screen should be:

Do you want to save these inputs?

(You can easily modify them later.)

1-Yes 2-No

Press 1 for yes. You will now be returned back to the Switch and Load Menu. We still need to set the percentage levels for the loads we just entered for switch station 01-5. From the Switch and Load Menu, press 1 for Examine/Update Switch Assignment. The next screen will prompt you to input the control station number. Press 01 and then press Enter at this screen. The output number screen will appear next. Press the 5 key and then press Enter. This screen will then be displayed:

VIEW - MODIFY/DELETE SWITCH ASSIGNMENT

1-View and modify switch assignment

2-Delete switch assignment

Press 1 to view and modify the switch. The next screen will display:

SWITCH 01-5 (ENTER for no change)

1-Change Type (Now a Scene Preset)

2-View/Change Assigned Load/Levels

3-Change Fade Time (Now 00 Min 02 Sec)

Press 2 to change the load levels for the kitchen path. This screen will appear next:

ASSIGNED LOAD LEVELS (ENTER-No Change)

01-3 : 50%          02-1 : 50%          02-4 : 50%

1-Delete          2-Add          3-More          4-Change Level

The default load level is 50%. We will leave load 02-1 and 02-4 at these defaults. They can be changed later at either the switch stations or the CCU. Load 01-3 is a non-dimming-type load that is controlled by a relay, so this needs to be changed to 100%. To do this, press the 4 key. The next screen will look like this:

CONTROL MODULE ADDRESS->_

Input the control module address and press ENTER.

Press the 0 and 1 keys, then press Enter. The next prompt will be:

OUTPUT NUMBER->_

Input the module output number and press ENTER.

Press the 3 key and press Enter. The next screen will look like this:

PRESET LEVEL? _

Input preset level (0 to 100 percent).

Then press ENTER.

Press 1 and then 00. Press the Enter key. The CCU will return to the assigned load display. Make sure that load 01-3 now has the value of 100%. If everything looks right, press the Back key until you are at the Switch and Load Menu. Switch and button 01-5 is now complete.

One more button will be completed in the same manner as the others. The rest of the switch stations can be programmed using the same basic methods.

From the Switch and Load Menu, press 1 for Examine/Update Switch Assignment. Input the control station address 01 and the switch position as 7 when prompted. Press 1 to assign the new switch and then select 7-Master from the Switch Type menu. This screen will then be displayed:

Master Type?

1-On 3-Toggle 5-Vacation

2-Off 4-Dimming

Press 2 for Off. This is the House Off button in the Foyer, so we will now need to assign all of the loads to switch button 01-7. The next screen will be the Control Module Address input. Press 01 and Enter. The next screen will be the Output Number, so press 1 and Enter. It will then ask if you would like to assign any more. Press 1 for yes. Continue inputting loads 01-2, 01-3, 01-4, 01-5, 01-6, 02-1, 02-2, 02-3, and finally 02-4 in the same manner. Once you have answered No for more inputs, save the inputs in the next screen by answering Yes. You will then be returned back to the Switch and Load Menu. Repeat the steps for switch station 01-9, but this time choose the On Master selection. When switch 01-9 is finished, switch station 01 in the Foyer will be completely programmed into the CCU.

Continue programming switch stations 02, 03, 04, 05, 06, 07, and 08. The Data Input Module FF will be programmed in the same manner, only choose 6-Timed/Flash from the Switch Type menu. The CCU will direct you to the screens necessary to program the 15-minute security on-time to input 1 (effectively button 1) of the Data Input module. When all of these switch stations are programmed, the basic programming will be complete.

There are many functions that the CCU can perform. In the earlier section on programming from the PC, an example of the CCU tests was already illustrated. Right now, we will go through some of the other features and functions of the Standard 2000 CCU starting with the second choice in the Switch and Load Menu.

### *Examining the Load Assignment*

SWITCH AND LOAD MENU

1-Examine/Update Switch Assignment

2-Examine Load Assignment

3-Unassigned Load Find

The second choice on this menu has not been selected yet. This option permits a way to check what switches are assigned to a particular load. You cannot make any changes to the data with this selection. When you press 2 from this menu, it will prompt you to input both the load address and output number. Let's say that you want to know what switches control dimmer module (02) load (2). When you input this load information , here is what you will see displayed:

02-2 is a Dimming Module - Output

Assigned Switches -> 01-7 01-9

05-2 08-7 08-9 FF-1

1-Quit (2-More)

If there are more switches, press 2. If not, press 1. In this case, these are all of the switches that control the dimmer module 02-2, so you would then be transferred to the Main Menu. Press 1 to transfer to the Switch and Load Menu.

The third choice on the Switch and Load Menu is 3-Unassigned Load Find. This option is used to find the address of an unknown load.

If a load has no control assigned to it, the CCU will locate and activate it. This is the display after you have selected this option from the menu.

Each unassigned load will cycle on and off. Press any station switch four times when load is on to stop the cycling.

Load Currently on -> 01-1 ENTER to quit

If the CCU finds an unassigned load, it will start turning that load on and off until you press a switch station button four times. The CCU will then display the load address that it was cycling. If no unassigned loads are found during this process, no load addresses will appear on the display and the Main Menu will return to the screen.

While we are still discussing features from the Switch and Load Menu, it is a good time to discuss the attributes of the some of the switch-type functions. Dimmer, Toggle, Master On, Master Off, Timed/ Flash, and Scene Preset have all been covered. There are still other button functions that were not used in this installation, so we will discuss them now.

The Master Dimmer function acts just like a regular dimmer, except that it controls more than one load in unison.

The Momentary activates its assigned load only when pressed. It can be used for applications such as motorized curtains or drapes. This function cannot be assigned to a Maintained Contact Data Input module.

The Master Toggle Off function will actually turn loads on and off. If all assigned loads are turned off, pressing this button will turn them back on. If all of the assigned loads are turned on, it will turn them off. If some of the assigned loads are turned on, pressing the button will turn them off. The LED for this switch will be off when all of the assigned loads are off.

The Raise function is used with a dimming load. When a button is programmed with this function and it is pressed, the load will brighten until the button is released.

The Lower function is also associated with a dimming-type load. It works in the opposite manner from the Raise function. It dims the load until the button is released.

The Open/Close function is used on motorized equipment such as drapes, curtains, skylights, projection screens, etc. It is designed to work with an Interlock Relay module. The Open or Close function is assigned in pairs to accommodate the forward and reverse leads of the motor being controlled. The pairs also associate to the interlock relay positions. Outputs 1 and 2 on the relay module are paired, as are 3 and 4. The pairing continues with outputs 5 and 6, and finally 7 and 8. On a six-channel interlock relay module, the pairing is as follows: (1 and 2), (4 and 5), and (3 and 6). This will conclude the discussion on functions. The next area that will be covered is setting the date and time.

### *Setting the Clock*

From the Main Menu you will see this display:

MAIN MENU (Free Memory -> 89%)
1-Switch and Load Menu  3-Set Clock
2-Timer Function Menu
-Back Space for normal operation_

We will select 3-Set Clock from this menu. The clock menu will then be displayed:

SET CLOCK MENU
1-Change Time
2-Change Date
3-Set Daylight Savings

Select 1-Change Time from this menu. The next screen displayed will prompt you to:

TIME (HOURS:MINUTES)? -> _
Input hours and press ENTER. Then
input minutes and press ENTER.

The menu is self-explanatory, so after the last Enter press you will be prompted to enter whether the entry means A.M. or P.M. After this selection, you will be returned to the main menu. At the main menu, press 3-Set Clock again to arrive back at the Set Clock Menu. From the clock menu, press 2-Change Date. The display will then look like this:

MONTH -> _
Input month (1-12; 1 = January) then
Press Enter.

Key in a number from 1 to 12 that corresponds with the current month. When you press Enter, this screen will appear:

DAY OF MONTH -> _
Input day (1-31) then press ENTER.

This screen is also self-explanatory. After the Enter is pressed, this screen will be displayed:

DAY OF WEEK -> _
Input day (1-7; 1 = Sunday) and
Then press ENTER.

Key in the current day that corresponds to the number identifier. After you press Enter, you will be transferred back to the main menu. Select 3-Set Clock Menu again. On the Set Clock Menu, there is only one option left: 3-Set Daylight Savings. Press 3 and this display will appear:

Daylight Savings is currently Inactive
Press 1 to toggle setting
Press Enter to save current setting.
Press back to abort

Pressing 1 changes Daylight Savings to Active. Pressing Enter saves the settings and returns you to the main menu. Pressing the Back key will return you to the main menu without saving. The clock is now set.

### *Timer Functions*

The main menu has one item that has not been covered yet: 2-Timer Function Menu. Press the 2 key to access it. This menu will then be displayed:

TIMER FUNCTIONS MENU:
1-Set New Time Flag
2-Examine/Modify Existing Time Flag
4-Vacation mode

The Timer Functions Menu allows you to program many different household controls. Timers are commonly thought of as simple devices that turn loads on and off at different times. Although this is true, these particular timer programs let you create some rather interesting approaches to house control. For example, in one installation the homeowners were having problems keeping their aerator pump to their fish pond turned on. They would have guests staying with them often and typically, someone would inadvertently turn the pump control button to off. This was a real problem because the rather expensive fish in that pond would die if that pump stayed off for too long. Using a timing function, the load was programmed to turn on at half-hour intervals 24 hours a day. It really did not matter if the load was already on when one of these timers activated; the CCU would send the command to the control module anyway. The maximum time that the aerator pump could ever be off would be 30 minutes using this program. This timing function gave the homeowner some much-needed peace of mind.

The same programming technique was used in another home. In this example, the homeowners used a set of wall lights that they dimmed manually to around the 20-percent level as their house night-lights. This manual dimming night-light arrangement worked well until their grandchildren came over to stay the night, which happened quite often. The children would get up and press a button that would turn the wall lights on to 100 percent. They would go to the kitchen or wherever and then go back to bed, leaving the wall lights on at the 100-percent level. The homeowners did not want these lights blinding them when they awoke, so they asked if something could be done about it. To repair this, 12 time flags were set to control a brand-new switch station location. The time flags were directed toward a switch that was programmed as a Scene Preset that would turn these lights on to 20 percent every half-hour throughout the night. Now, every time the lights are turned on to their full brightness during the night, the time flags instruct the switch to turn the wall lights back on to 20 percent. This procedure effectively dims them back down.

The strange part of this is that the time flags directed these commands to a switch that did not physically exist. This sounds impossible, but you can program a switch station into the CCU that does not actually exist on the switch station bus. There is a good reason to create a program like this. The existing house switch stations did not have a spare button with which to program a new 20-percent Scene Preset. With regard to this problem, the decision was made to create a hypothetical switch. The next step was making sure that an existing switch station address was not reused in this procedure. A switch station address was decided upon, then it was programmed with the function of Scene Preset with a load dimming level of 20 percent. This switch will never be found by using a control station test, at least not physically. However, the CCU will look to the load/function information contained in this hypothetical switch, then it will turn on the real load to 20 percent. You can see by these examples that some timing functions are really only limited by the imagination and necessity.

We will now access the first selection in the Timer Functions Menu by pressing the 1 key. This screen will be displayed:

TIMER TYPE:
1-Daily/Weekly
2-Specific Date (Yearly)

Selecting 1-Daily/Weekly will allow you to enter the day and time that you would like the switch to be activated. Selecting 2-Specific Date (Yearly) will allow you set a date during the year for the switch to be activated. The CCU will lead you through the steps to accomplish both of these tasks. The first screen could be either one of these (depending on which selection you choose):

Toggle Selected days of the week by
Pressing keys 1-7 (1 = Sunday) and
then press ENTER.
sun mon tue wed thu fri sat
or this screen:
MONTH ->_
Input month (1-12); 1 = January) then
press ENTER.

After these inputs are made, the CCU will prompt you to enter the time of day. The time entry is the same for both selections. The screen will look like this:

TIME TO BE SET IS:
1-Clock Time
2-Relative to Sunrise
3-Relative to Sunset

Choosing the first option will then display this prompt:

TIME (HOURS:MINUTES)? -> _
Input hours and press ENTER. Then
Input minutes and press ENTER.

After the time is entered in, you will be prompted to input:

TIME (HOURS:MINUTES)? -> 6:00
Press 1 - A.M.
2 - P.M.

Press either the 1 or 2 keys. The next screen will prompt:

CONTROL STATION ADDRESS -> _
input station address and press ENTER or
just press ENTER for switch Identify.

If the switch station address is known, enter it now. The following screen will then appear if you entered the address:

SWITCH POSITION NUMBER ->_
:1 2 3 :
:4 5 6:
:7 8 9:

Press the button number that you would like timed. After the input, the next screen that will be displayed will prompt you:

Do you wish to assign more?
Press 1-Yes 2-No

Press 2 for No and the following prompt will appear:

Do you want the loads controlled by
Switch 01-1 turned on or off?
Press 1-On 2-Off

Press the appropriate key and a review will be displayed:

You have requested a time flag for
DATE Jan 11
TIME 6:00 PM
Is this correct? 1-Yes 2-No

Press 1 for yes and the CCU will take you to the Main Menu. Setting a new time flag can also be accomplished by using the astronomical clock function. After choosing the Timer Type, which sets the flag time and date, select either 2-Relative to Sunrise or 3- Relative to Sunset. These options will permit selected switch/load activations to be more accurate in regard to the actual periods of darkness.

From the main menu, press 2-Timer Function Menu. The next option we will inspect is 2-Examine/Modify Existing Time Flag. This is a review-type selection, but you can make changes to the time flags that have already been set. This is the menu that will appear when you press this selection:

SELECT TIMER TYPE (FLAGS SET)
1-Daily/Weekly (001)
2-Specific Date (000)

This menu will show how many time flags have been set for each selection. Pressing the 1 key will allow you to view and modify the flags that have been set. The next screen will look like this:

Daily/Weekly Time Flag 01 of 01
SmTWtFS 06:00 PM
1-1 On
1-Modify 2-Next 3-Quit (4-more)

Pressing the 1 key will allow you to modify the selection. The next screen will then display:

FLAG MODIFY:
1-Change Type, Time/Day/Date
2-Change assigned Switches, On/Off
3-Delete Flag

Choosing the third item will delete the current time flag. When you pack the CCU memory on some later occasion, this time flag deletion would then be totally purged from the CCU. Pressing either the 1 or 2 keys will allow these screens to appear:

MODIFY TYPE or TIME:

1-Change Type

2-Change (Date or Day and) Time

Or:

SWITCH UPDATE

1-Add a Switch 4-Done

2-Delete a Switch

3-Change On/Off

Choosing selections from either of these two screens will allow you to make the necessary changes to the time flag. During this modifying period, all of the screens that you come in contact with will be familiar to you from your earlier time flag programming. When you have completed each change procedure, the CCU will return you to the main menu.

From the main menu, select 2-Timer Function Menu once more. The timer function menu has one more option that has not been chosen yet: 3-Vacation Mode Menu. In this menu, the CCU will guide you through the same procedure that you went through when you set a new time flag. The only difference is that in the random vacation mode, the program will require both an On and an Off time to be set. If you choose to enable the random vacation mode, all other programmed time flags will then be disabled. Care should be taken when choosing this option because some time flags may be critical to normal house control.

It should be noted that many earlier CCU versions that are still in use may have a different time flag programming environment. In some ways, the earlier versions are easier to work with and have some of the same characteristics that the newer versions have. Adjusting between CCU time flag environments should not be a major obstacle once one of the versions is learned. The timer functions have now been completed. The next section will cover the important aspects of the Options Menu.

### Options Menu

This menu is mainly used to set certain parameters in the CCU environment. It is generally accessed when the CCU is first installed and tested. After that, access normally is only necessary when a change or a problem occurs within the Standard 2000 system. This menu is accessed by pressing the Option key on the front of the CCU while the normal operation screen is being displayed (the one displaying the current time and date). It is important to the normal operation of the system that you back out of the options menu when you are finished using it. Parts of this menu were discussed earlier in the uploading to the CCU and system tests sections, but we will discuss them again in more detail. This is the screen that will be displayed once the Option key is pressed:

OPTIONS MENU
1-System Parameters 4-Pack Memory
2-System Tests 5-Diagnostics
3-LiteTouch Phone Numbers

### System Parameters

Pressing the 1 key will access the system parameters menu. There are six options to choose from in this menu:

SYSTEM PARAMETERS
1-LED Invert 4-Switch Options
2-Computer 5-Poll Setting
3-Telephone 6-View Lat/Longitude

Pressing the first selection will cause the following screen to appear:

SWITCH TYPE
1-Dimmer 2-Open or Close
3-Raise/Lower 4-Toggle 5-Momentary
6-Timed/Flash 7-Master 9-Scene Preset

Choosing any of these options allows you to change the current LED status of each. If the button LEDs of a certain function turn on when the load turns on, they are noninverted. If they act in an opposite-type manner, they are considered inverted. Here is an example of the screen you will see if you press 1-Dimmer:

Dimmer switch LEDs are

[noninverted]

1-Change status

2-Exit without change

Pressing the 1 key will change the noninverted dimmer LED status to inverted. Pressing the second selection will return you to the options menu. You can choose to change any or all functions in the switch-type menu in reference to their LED status. One common reason that some homeowners prefer to reverse the LED status is that it makes the buttons easy to find in the dark. On the older LiteTouch systems this was a common change. This status change is not as prevalent on the newer systems, however. The older switch station LEDs were totally dark when all of the lights were off in the house. The newer switch station LEDs are partially lit in the dark. The LED level of an off switch versus an on switch is quite different, so usually a mistake in misreading the LED status does not occur. After a change in the LED status is accomplished at the CCU, the switch stations involved with this change will have to be used to implement this modification (push one of the buttons on the switch). When you are completed with the changes in the LED status, you will then be returned to the options menu.

Press 1-System Parameters at the options menu again. This will bring you to the System Parameters Menu. The next item on this menu is 2-Computer. This was one of the items covered previously in the uploading to the CCU section. Pressing 2 will allow this menu to appear:

Computer Interface Option Menu

1-Change Communications Settings

2-Change Extended IR Command Settings

Press the 1 key and you will see this display:

Select Desired Baud Rate

1 – 9600      4 – 1200

2 – 4800      5 –  600

3 – 2400      6 –  300

Press the 1 key for LiteTouch software. The baud rate of 9600 is used on the LiteTouch Standard 2000 system. After the baud rate is selected, the next screen will appear:

Select Parity/Stop Bits

| | |
|---|---|
| 1 – None/1 | 4 – Even/2 |
| 2 – None/2 | 5 – Odd/1 |
| 3 – Even/1 | 6 – Odd/2 |

Press 1 for LiteTouch software. The LiteTouch Standard 2000 system uses no parity and one stop bit. The second choice in the computer interface option menu is 2-Change Extended IR command settings. You may see this option in the CCU you are working on. LiteTouch no longer provides support for this function. Please contact the company for any further information concerning this. After you have completed the computer interface section, the CCU will return you to the options menu. Press 1 from the options menu to access the system parameters menu again.

The next choice in the system parameters menu is 3-Telephone. This function allows a new security code to be entered along with the number of rings until a CCU answer. Here is the menu that appears when this option is selected:

Telephone Interface Option Menu
1-View/Modify Security Code
2-Change Answer Ring Number (now 4)
3-Toggle Answer Mode (Now Voice)

Pressing the 1 key will show either of these two screens:

Telephone Security Code Inactive
Present Code -> 123456789
1-Activate Security Code
2-Change Code

or this screen:

Telephone Security Code Active
Present Code -> 123456789
1-Inactivate Security Code
2-Change Code

Pressing the 1 key will either enable or disable the security code. Pressing the 2 key will allow this prompt to appear:

Input new code (up to ten digits) then
Press ENTER.

—

If the security code number has been activated, this code number will have to be input from a touch-tone telephone or PC modem to allow off-site telephone CCU access. Once the new code has been entered into this field, the CCU will return you to the options menu again. Press the number 1 selection again to bring you back to the system parameters menu.

Press the 3 key again from the system parameters menu to bring you back to the Telephone Interface Option Menu. The second choice in this menu is to change the number of rings. Press the 2 key and this prompting screen will appear:

On what ring (1-9) do you want
LiteTouch to answer the phone?
Input number then press ENTER.

—

After a number has been input, press the Enter key and you will be returned to the options menu again. Press the 1 key from the options menu and the 3 key from the system parameters menu. This will bring you back to the Telephone Interface Option Menu. The last selection on the list will permit you to either select voice or modem capability. The voice capability will allow a touch-tone telephone to control the house functions. When a call is placed to the LiteTouch telephone line, a voice will answer that says, "LiteTouch 2000." This will mean that the system is online. If you have a security code activated, the caller must input it at this time. If the wrong code is entered, the connection will be broken. If the code is entered correctly it will say, "Ready." The caller will input the * key on his or her telephone, followed by the three-digit switch station address that has been selected to control (for instance, pressing *01-1 will turn on the outside light in this installation). The CCU will respond by saying, "Switch 01-1 on." If you wish to turn any switches off, press the # key and the three-digit switch address (#01-1 would turn the outside light off). The CCU will respond, "Switch 01-1 off." If a switch that is not programmed is entered, the CCU will reply, "Switch XXX invalid." When 30 seconds pass without any touch-tone keys being pressed, the CCU will break the connection. If you are finished with your CCU inputs, hang up. No special commands are needed to end the CCU telephone session.

When the modem capability is enabled (versus voice), this will allow a PC to perform diagnostics and upload/download support of the LiteTouch system.

After the telephone interface has been properly set up, you will be returned to the options menu again. The next selection in this menu is 4-Switch Options. Press the 4 key and this menu will appear:

SWITCH OPTIONS MENU

1-Dimmer Options 4-Interlock Type

2-Scene Options

3-Timed Switch Toggle Setting

This is the menu that will appear when you press the 1 key from the switch options menu:

DIMMER OPTIONS MENU

1-Set Dimmer Ramp Time

2-Set Minimum Turn On Level

Pressing the 1 key allows you to change the speed at which the dimmer function controls a dimming load. The default ramp time is 3 seconds, but you can set it anywhere between 2 and 6 seconds. Pressing the number 2 key will allow you to set the lowest dimmer load activation level. You may have noticed that on standard household dimmers, there is a setting where no visible light is emitted from the load that it controls. Usually this is right after the turn-on click on some types of dimmers. The minimum CCU turn-on level setting deals with this effect. You will set a minimum level so that this "dead" area can be eliminated. The factory sets this amount to 25 percent. This percentage is called the minimum default turn-on level, but it can be set to a lower amount if you desire. If you press the number 2 key from the menu, the following will be displayed:

The current dimmer minimum turn-on
level is 25 percent.
Input new minimum level (0-50)-> _

Press the front CCU keys that correspond to the level you desire and then press Enter.

The next selection in the Switch Options Menu is 2-Scene Options. This is the program area that allows you to lock or unlock the scene preset buttons throughout the house. When this selection is set in the unlock mode, the homeowner can effect changes to a scene preset without any programming knowledge. For example, a scene preset button has eight loads assigned to it. In addition, these loads also have eight individual controls (buttons) located in different switch stations around the house. The homeowner will go to each of these eight buttons and dim their individual loads to a desired level. He may even decide to leave one button completely off because he does not wish that load to come on with the current scene. When these have been completed, the homeowner walks back to the scene preset button that controls all eight loads and holds it for at least 5 seconds. The scene preset button will now mimic the settings that the homeowner has just set. This is a great programming option. The only problem with this is that sometimes the system is not returned back to its locked state. The next time someone holds any scene button in for more than 5 seconds, the scene button being held will mimic the undesirable load setting. You can imagine how this will cause the overall system to appear like it has a serious problem, when in fact it does not. When you view/modify the scene switch station address, you can tell when something like this has happened. All of the loads of the scene are set at 00%. This zero percentage is a telltale sign that someone has pushed too long on a scene button with the system unlocked. Reset the scene values back to their original settings and lock the system. Here is how you can do this. Press 2-Scene Options from the switch options menu. This screen will then appear:

SCENE OPTIONS MENU
1-Lock/Unlock
2-Toggle Setting

Press the 1 key. The next screen will look like this:

Scene Preset Levels are now unlocked
They can be altered at the Control
Stations.
1-lock levels 2-exit

Press the 1 key to lock the scene preset levels. If you want to unlock them again, select this same screen and the CCU will then give you the choice to unlock them. It toggles the choice to the opposite setting. The second selection in the scene options menu is to change the scene presets to toggle or not to toggle. You can make a choice and then press 2 to exit to the system parameters menu.

From the system parameters menu, press the 4 key to return back to the Switch Options Menu. The next option on this menu is 3-Timed Switch Toggle Setting. Press the 3 key and one of the following menus will appear:

Timed Switches toggle.
1-Change Toggle Setting
2-Exit

Or:

Timed Switches do not toggle.
1-Change Toggle Setting
2-Exit

Pressing the 1 key on either selection will change the setting (toggle means on and off). When your choice is complete, press the 2 key to exit. You will then be returned to the system parameters menu.

Press the 4 key from the system parameters menu to bring you back to the switch options menu. The last item in this menu is 4-Interlock Type. Press the 4 key and one of the following two menus will appear:

Interlock Relay Type Used
(currently 6 output/ 3 interlock)
1-Change Interlock Relay Type
2-Exit

Or:

Interlock Relay Type Used
(currently 8 output/ 4 interlock)
1-Change Interlock Relay Type
2-Exit

As was mentioned before, the earlier relay control modules were a six-channel (output) type. These are still available for replacement reasons. The eight-channel has become more prevalent in recent years, however. It is important that the two different interlock type relays are not mixed together in the same system. The above setting requires that either a six-channel or an eight-channel interlock relay be selected. This does not mean that you cannot have a standard six-channel relay used in a house that has some eight-channel relays already installed. It only means that you cannot use both relay types for interlock (motor control) reasons. Press the 1 key to change the interlock type relay that is used in your particular installation. Press the 2 key to exit to the system parameters menu when you are through.

The next item on the menu is 5-Poll Setting. Press the 5 key to turn switch station polling on or off. In this system, the CCU interrogates the switch stations as to their LED status. This is a periodic routine check when this option is enabled. Polling is only a system precaution, so it is not critical that this setting is activated. If for any reason your control stations do not have LEDs (this is uncommon), this setting should be selected to off. If you have a very large system and you feel that the CCU is responding too slowly, it wouldn't hurt to disable this feature. It may cut down on some of the CCU activity. You should be returned back to the system parameters menu once you have selected the poll setting.

The last item on the system parameters menu is 6-View Lat/Longitude. Press 6 to inspect the degrees and minutes of the latitude and longitude set into the CCU program. This information is used for the system's astronomical clock, which can accurately control nighttime or daylight time flags. Currently, this setting cannot be adjusted at the CCU. A PC upload containing the correct astronomical information must be used to make these changes.

Press the Back key until you arrive at the Options Menu. The next selection on this menu is 2-System Tests. Press the 2 key and this screen will appear:

SYSTEM TESTS MENU

1-Toggle Switch Display Flag (Now Off)

2-Module Communication Test

3-Station Communication Test

This testing menu has been mentioned previously in the System Tests section (CCU install). It wouldn't hurt to go over a few of the points again. Pressing the 1 key will turn the toggle switch display flag on. Pressing the 1 key again will turn it off. When it is activated, pressing a switch station button will allow the address of that button to be displayed at the CCU. It is important that you toggle this option to the off mode when you are through using it. If it is inadvertently left in the on mode, the switch station LEDs will not update (turn on with the load). This feature disables the normal LED status indicators.

The second selection is 2-Module Communication Test. Pressing the 2 key will make this test screen appear:

Module Communication Test

Press any key to end after current loop

Loop 01 Address 00

During loop 01, the address will increment from 00 to 99. After address 99 is reached, loop 02 will then start the address increment again. If no control modules are found after loop 01, the program will stop and display this message:

No Control Modules Found

Press any key to go on.

If any control modules are found, the test will continue until loop 99 is reached. It will then give a report of the test. Here is an example of a test result:

Test Results (press a key to continue):

01 Reported Every Loop

02 Missed Reporting on 03 loop (s).

03 Programmed but not reporting

In this test result, control module 01 has good communications with the CCU. Module address 02 probably has a bad connection or a broken wire at its data-in port or at the data-out port of the previous module. Control module 03 is not communicating at all. This may simply be a control module circuit breaker or fuse causing this error. It could be that this module does not even exist, but an old control module program is still residing in the CCU. It could mean many different things. The section on troubleshooting will address more of these issues. Pressing Enter after this test is complete will return you to the install tests menu (in some software versions this will be the System Tests Menu). From the system tests menu, select 3-Station Communication Test. This test uses five different steps to accomplish its task:

1 - Each control station port is tested for basic integrity.

2 - All LEDs on all control stations are turned off.

3 - A response is requested from all switch control stations.

4 - All LEDs on all control stations are then turned on.

5 - Another response is requested from all switch control stations.

Make sure that the switch station buttons are not pressed during this test. Button presses will be logged as errors. It will be difficult to verify a switch problem if the test results are not accurate. When the control station test is activated, this screen will then be displayed:

The System will be inoperative during
the control station communication tests.
Switch presses will be logged as errors.
Press any key to go on.

After you have pressed a key (other than Cancel), the next screen will look like this:

Pressing CANCEL will abort station test.
Testing CCU port controllers

If a problem arises from this port test, the following screen will then be displayed:

Pressing CANCEL will abort station test.
Testing CCU port controllers
Port 1 is Non-Functional
Press any key to go on.

This port trouble could be linked to a couple of different problems. Most likely, one of the switch stations is bad and is causing the port to shut down. The problem could also be a bad port. You can check this by pulling the orange port connector from the back of the CCU and retrying the test. If the problem persists, the trouble is most likely in the CCU port. These are the screen-displayed port numbers with their corresponding identifiers referencing the switch station connectors on the back of the CCU: Port 1 - TB5, Port 2 - TB6, Port 3 - TB7, Port 4 - TB8.

When the port testing is completed, the CCU then turns off all of the switch station LEDs. This message will be displayed:

Pressing CANCEL will abort station test.
Turning off all switch LEDs
address-> 00

The switch station address will progress from 00 to FF. After all of the LEDs are turned off, the CCU completes a switch station search. This screen will be displayed at this time:

Pressing CANCEL will abort station test.
Searching for Control Stations
address-> 00

There are two error messages that can result from this test. One is Station Reporting out of order and the other is Station Reporting for the second time. Stations reporting out of order may mean that someone has pressed a button during the test procedure. It can also mean that the switch station line may be experiencing interference problems. This may require pull-up resistors to be added to the data line. LiteTouch recommends that you call it before attempting this procedure.

A control station reporting for a second time may mean that two switch stations have the same address dialed into their most significant and least significant digits. This can only be detected by the CCU if these two like addresses are located on different ports. This kind of problem will cause switch station LED update trouble. The testing procedure will wait for you to press Enter before continuing the test when either of these two problems is detected.

When the station search has been completed, the LEDs are then turned back on. The following screen is then displayed:

Pressing CANCEL will abort station test.

Turning on all switch LEDs

address-> 00

The address will progress from 00 to FF. The CCU will then complete another search after the LEDs are turned back on. This is the next screen that will be displayed:

Pressing CANCEL will abort station test.

Searching for control stations

address-> 00

The station addresses will progress from 00 to FF. There are three possible error messages that may appear at this time. The first two you already know: Reporting out of order and Station Reporting for the second time. The third error message is Station not reporting with LEDs on. This error can point to a couple of different system problems. One is that a power supply is not operating (faulty unit, blown breaker or fuse). Another possible reason is that the switch station line is power anemic and may need an additional power supply added to the switch station bus. LiteTouch recommends that you consult with it before adding another power supply to the switch station wiring. When the switch station test is complete, the LEDs will be left on. If you desire, you can run the switch station test again and cancel the test at the part that turns off all of the LEDs. Pressing some of the switch station buttons after this test has been completed will permit the current LED updates to occur.

At the Options Menu, the next selection is 3-LiteTouch Phone Numbers. Pressing this key will provide you with the current telephone numbers that you may need for support.

The next selection is 4-Pack Memory. Pressing the 4 key will allow this function to be activated. The following screen will then appear:

Packing Memory

Removing Deleted Time Flags

Packing Switch/Load Area: XXXX DDDD

Each line on this screen will appear as it accomplishes its specific task. Packing the memory will permit the CCU to operate more efficiently because it reallocates the old unusable memory. This process will take longer than deleting the old time flags. The packing switch/ load area will display two four-digit hexadecimal numbers. The XXXX is the ending address of the pack. The second number, represented by DDDD, is the current location in the packing process. This number will slowly progress until both numbers attain an equal value. The LiteTouch system will act sluggishly during this packing process. After this process is complete (about 20 minutes), you will be returned to the options menu.

From the options menu, select the last item, which is 5-Diagnostics. Pressing the 5 key will allow this menu to appear:

Diagnostic Menu

1-Run Switch/Load Definition Check

2-Reset CCU without clearing RAM

3-Clear RAM and reset CCU

Pressing the 1 key will complete a memory check of the switch and load definitions. Pressing the 2 key will reboot the CCU and allow the RAM to remain intact. Pressing the 3 key will clear the RAM and reset the CCU to the factory defaults. It will then complete the same operation that pressing the 2 key accomplishes. The RAM contains all of the current switch and load settings, so warn the homeowners that the house will lose the current switch settings when you clear it. The programmed switch and load address information will still be present after the RAM is cleared. It will not wipe out this important information.

This concludes the chapter on the programming of the Standard 2000 CCU. Creative programming is one of the keys to building an interesting and useful automation system. If you worry about an unusual programming technique that you may want to use, just ask LiteTouch support about it. The company will help you make the correct decision. In this type of system, if you can imagine a way to program a piece of automation equipment, most likely it can be accomplished. Correct programming and common-sense interfacing are the main contributors to a successful automated outcome.

# 4

# Installing and Programming the 5000LC CCU

## About the 5000LC

The 5000LC controller (Figure 4-1) is very similar to the Standard 2000 CCU in that it receives digital input from the switch stations and sends the corresponding data to the control modules. The same switch control stations and control modules that are used with the 2000 system can be used with the 5000LC CCU. The CCU is designed to expand to the future needs of home automation by the implementation of separate slide-in cards. The cards that are available now are able to monitor load and switch status along with providing surge suppression capabilities. In addition, they also handle modem, CPU, computer interfacing, and many troubleshooting capabilities. These are only some of the complex duties that these cards can perform.

The 5000LC is programmed exclusively from an external PC using the LiteWare design and program software. Programming from the LiteWare environment will be covered later in this chapter. One way that the 5000LC CCU differs from the Standard 2000 is that it does not have an LCD screen, although a separate computer display can be attached to the 5000LC system. A separate PC resident upload/download software called CP5000 is used to diagnose and maintain the system in an effective manner.

The 5000LC CCU has many integral parts that comprise the entire controller system. They include a separate enclosure, power supply,

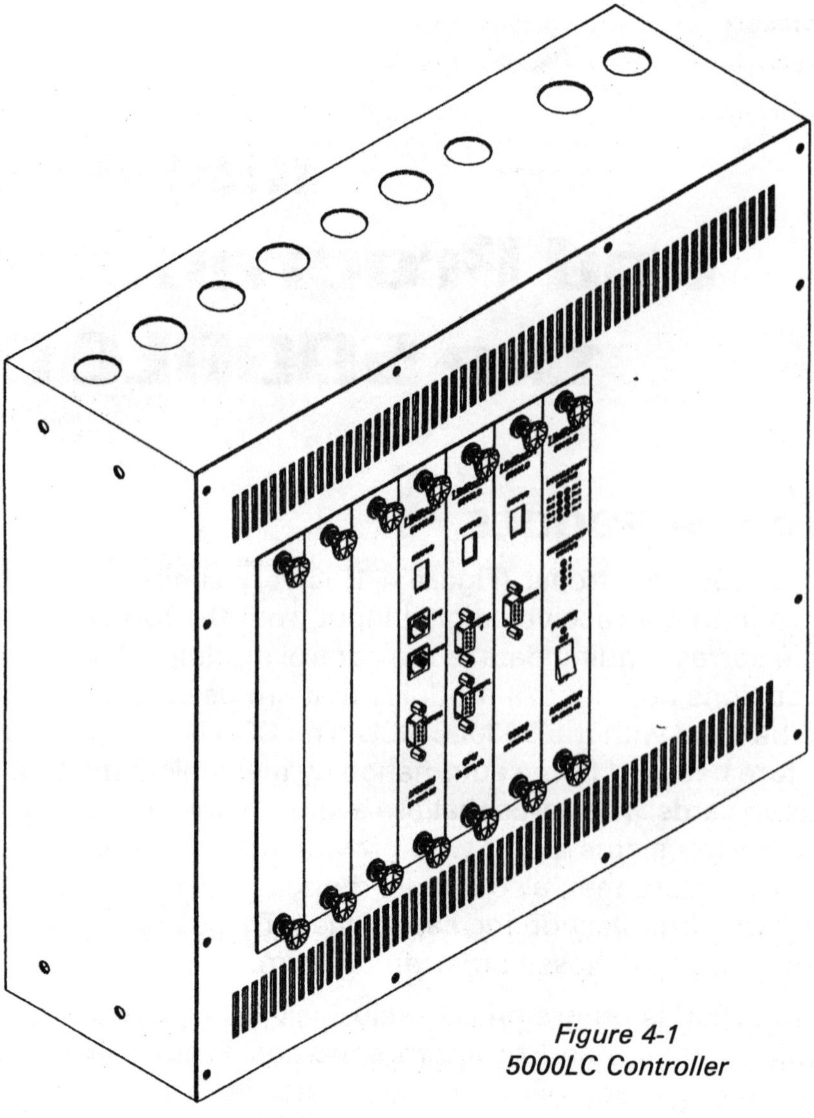

*Figure 4-1*
*5000LC Controller*

back plane board, and numerous individual slide-in cards. The CCU wiring is very similar to that of the Standard 2000 system, so a retrofit from the standard to the 5000LC is usually not a major problem. The next section in this chapter will illustrate a simple retrofit. In fact, the same installation that was used earlier to demonstrate the Standard 2000 system install will be used as a retrofit example.

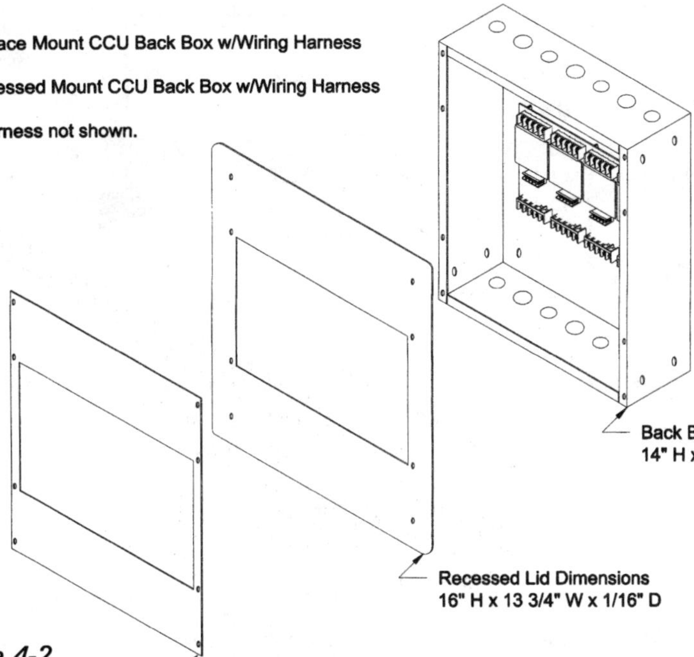


*Figure 4-2*
*Back Box*

The 5000LC comes with hardware and information packages. One package is usually for the rough-in stage of the installation and one package is for the finish stage. Here is a hardware list for a typical installation (this is considering that the modules, module enclosures, switches, and cabling are already retained):

- Back Box: (Figure 4-2) This rough-in enclosure holds the components, isolates the high- from the low-voltage and routes the switch station and control module cables.
- Lid: (Figure 4-2) The lid can either be used for surface or recessed mounting. In the recessed configuration, it will overlap the back box edge about an inch all the way around. It is used to cover the high-voltage terminations and also helps the system cards slide back to connect with the back plane board.
- Back Plane Board: (Part of this board can be seen in the rear of the back box in Figure 4-2. A detailed back plane figure will be used later

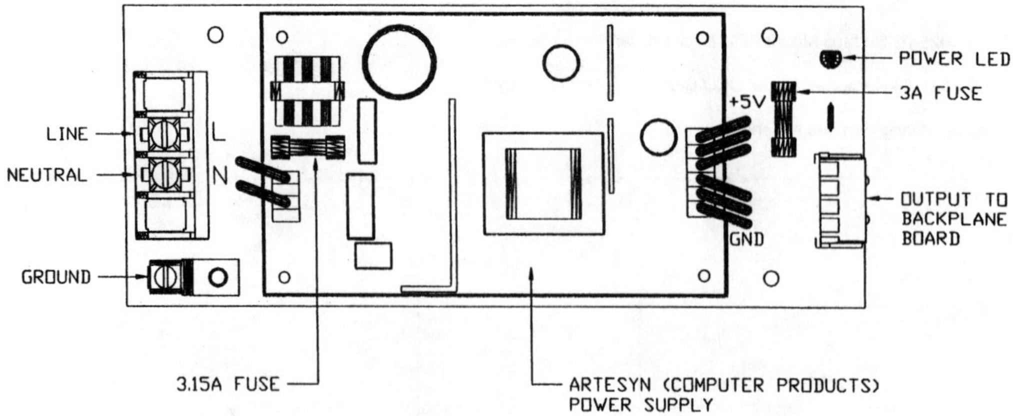


*Figure 4-3*
*Power Supply Board*

in this section.) The back plane board is where the control module communication cables and the control station bus are terminated. In addition, the system cards are slid into the connections on this board.

- Power Supply Board: (Figure 4-3) This board supplies the power to the back plane board and system cards. It is located in the right-hand partition area of the back box where it will be supplied with a 12-2 110-volt ac home-run from the electrical distribution panel. A special wiring harness will run from this power supply board to the back plane board located in the left-hand partition of the back box.
- Cable: This is the special wiring harness that connects the power supply board to the back plane board.
- Dust Cover: This cover protects the inside of the enclosure from dust and debris during the construction phase.

Here is a list of enclosed information that will accompany the 5000LC hardware package:

- 5000LC Installation Guide: This booklet will describe how to properly wire the 5000LC to the control modules and switch stations.
- 5000LC User's Guide: This booklet will describe how to install and troubleshoot the 5000LC hardware.
- Power Bus Monitor: This tool will indicate communications activity between the 5000LC controller and the control modules.

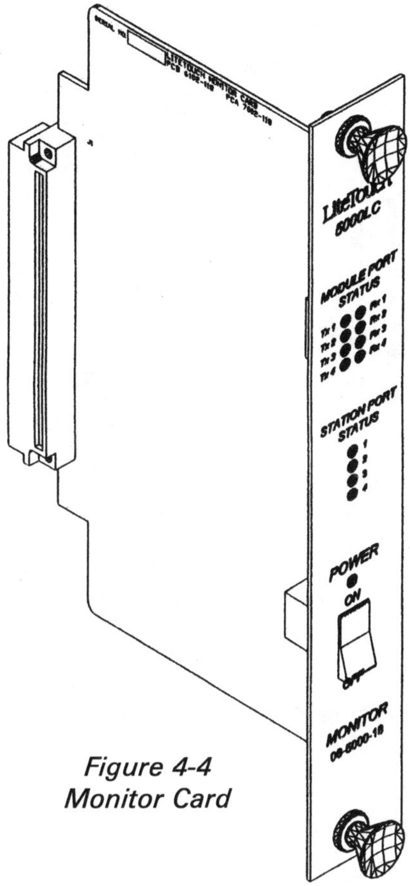


*Figure 4-4*
*Monitor Card*

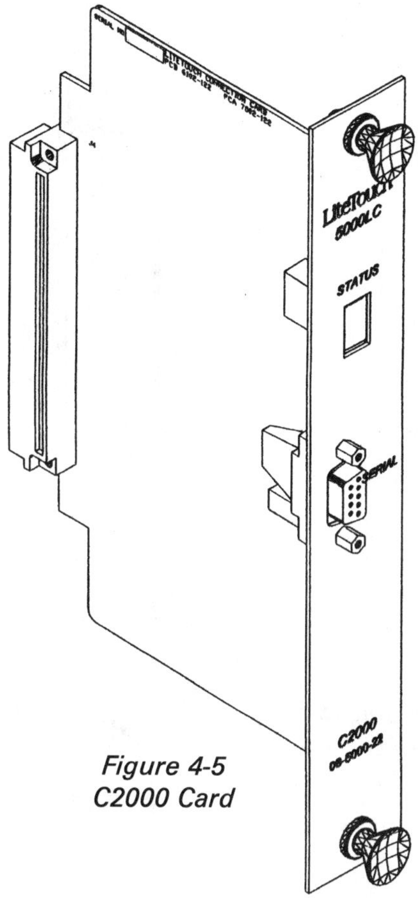


*Figure 4-5*
*C2000 Card*

- Warranty Information: This will inform you about returning equipment and details of the warranty.

The next list of card hardware and information items are usually acquired during the finish part of the installation:

- Monitor Card: (Figure 4-4) This card must be installed in the right-most slot on the back plane board. It switches power to the rest of the system cards, and it also monitors station and module bus communications.
- C2000 Card: (Figure 4-5) This card communicates with the 2000 system control modules and switch stations. It can also act as a backup in case the main central processing unit fails.

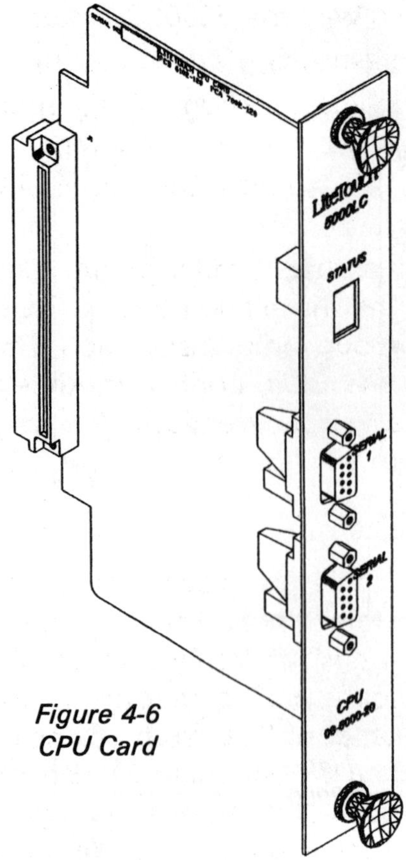


*Figure 4-6*
*CPU Card*

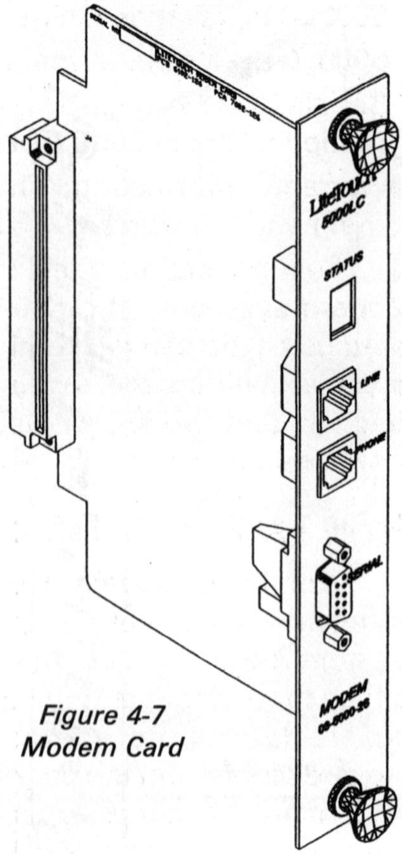


*Figure 4-7*
*Modem Card*

- CPU Card: (Figure 4-6) The central processing unit card is the main brain of the 5000LC system. It controls all of the system operations and also retains the job programming information in its memory.
- Modem Card: (Figure 4-7) This provides the telephone interface capabilities to the entire system. It can be used to control the 5000LC system remotely.

Along with the card hardware package, this information to complete the system is also included:

- 5000LC Programming Manual: Shows how to use the programming software.
- 5000LC Programming Software: This is the LiteWare program that will be discussed next.

- 5000LC Installation Guide: Shows how to wire the 5000LC system.
- 5000LC User's Guide: Installing and troubleshooting is described here.
- Serial Cable: A nine-pin cable used to upload and download information to the 5000LC CCU from a separate personal computer.
- Warranty Information: This will inform you about returning equipment and the warranty details.

The next section of this chapter will cover the retrofit from the 2000 standard system to the 5000LC system. The installation will be illustrated using the same specifications used in our earlier installation. The floor plan will be the same. In addition, the CCU, control modules, switches, and module enclosure locations will be identical. You may want to page back to this installation area to use it as a reference.

## How to Retrofit the 5000LC

All of the work that will be accomplished will take place in the utility room and at the telephone interface box on this retrofit. The first important item is to make sure that all power is turned off. Verify with a high-voltage meter that the three 20-amp feeds to the control module enclosure are de-energized. Remove the 2000 standard CCU from its enclosure and disconnect the connectors located on the back side. Mark each switch station cable as to its grouping/purpose and remove the pigtail wire nuts connecting this cable group together. Mark and then remove the power supply cable wires from the orange connector. Loosen the rest of the setscrews on the orange connector and set the pigtail wires aside. Then, loosen the setscrews on the control module communication connector and remove the cable. Since this is the only communication cable in this box, you do not need to mark it.

Disconnect the telephone cable at the outside interface (this line has the possibility of being shorted during this retrofit). Then, disconnect the telephone line from the CCU connector. Remove the grounding wire from the CCU and carefully place the 2000 CCU in a protected area away from where you are working. The next step will be to remove the 2000 enclosure. Most likely the enclosure has been screwed into the studs on the left and right side of the can. Remove these screws and carefully pull the enclosure out of the wall. The enclosure connectors or the cables themselves may give some resistance. Try pushing the cables out from the inside of the can so that they will not resist the can movement as it is pulled from the opening. If clamp-type cable connectors were used, try to loosen them when the top of the can appears outside of the opening. This will permit the cables to slide out more freely. After a bit, the can should be able to be removed totally and set aside. Place the cables back inside the opening so that they will not be in the way for the next step. Locate the 5000LC back box and compare its size to the existing drywall opening. Place it over the opening and level it. Using a pencil, trace the outside of the box. Use a level to check the pencil markings afterward to make sure that the can did not move when you made the new cutout marks. There may be instances where areas of the existing opening are larger than the back box that you are installing. In this situation, measure the difference and see if the 1-inch overlap of the lid will cover it. If it doesn't, some spackling may have to be used later to hide this difference. Now, using a drywall saw or a utility knife, carefully cut the pencil-marked areas away. Then, check to see if the back box will slide into the adjusted opening. There may be a need to remove or adjust the stud-work inside the wall. This can be difficult to do without harming the surrounding drywall. As long as the wall is not a load-bearing-type, most minor stud adjustments will not weaken the structural integrity of the wall. When the back box fits flush into the new opening, proceed to the next step.

If you are not an experienced electrician, you may need to call one for the next step. A separate 12-2 Romex cable needs to be run from the distribution panel to this new opening. The distribution panel will be energized even when the main breaker to the house is off, so care needs to be taken when removing this high-voltage panel cover. Ask the electrician to leave about 2 feet of extra cable protruding out of the new back box opening. Make sure that the 12-2 cable has not been connected to the high-voltage panel yet. This will need to be accomplished when the hardware retrofit is nearly complete and when there is less of a chance that you or anyone else will come in contact with the high-voltage leads. Once the de-energized 12-2 cable is run, we can move on to the next step.

The modem card on the 5000LC has two RJ11 female jacks located on its front side. This means that the existing telephone cable cannot be inserted inside the new back box. It must be run to a new box and terminated to another RJ11 female jack (just like a standard house telephone jack). If the telephone cable was previously run overhead and then brought down to the CCU, the telephone box will probably need to be installed above the new CCU. This will be easier to accomplish when the back box is not installed. Using a one-gang cut-in box, trace the outline of it at a point above the CCU opening. Once the opening is cut to the size of the box, feed the telephone cable up through the CCU opening to the one-gang box opening. Slide the cable end into the cut-in box and then fit the box into the new opening. Secure the cut-in box and strip some of the outer sheathing off of the telephone cable. Wrap the unused conductors back and then attach the stripped wires to the RJ11 cover plate assembly. Match the red/green wire terminals on the RJ11 to the red/green wires of the cable and then secure. Then, screw the RJ11 cover plate assembly to the box. When the retrofit is complete, a double-ended male RJ11 telephone cord will be connected from this new box to the modem card on the front of the 5000LC. The connections at the telephone interface can now be reconnected.

It is now time to install the cable connectors into the back box. When this step is complete, start feeding the existing cables into the connectors. The separate grounding wire for the 2000 system will not be needed for this CCU; it will receive its grounding through the high-voltage feed. Make sure that the 12-2 Romex is the only cable entering the high-voltage right-hand side of the partition. Without pinching the cables, slide the back box into the wall. Pull any excess cabling that may have gathered inside the wall down into the back box. You will probably need as much switch station cable as you can gather since these connections will be located near the bottom of the back plane board. Secure the back box to the studs making sure that the box is both level and flush with the wall. Ask the electrician to connect the 12-2 cable to the power supply board (Figure 4-3, line, neutral and ground), but ask him not to connect the other end to the electrical panel yet. The rest of the terminations will be covered next.

The power supply, switch station, and communication cables are terminated exactly in the same order as was the 2000 CCU (Figure 4-8, back plane wiring). The power supply positive is terminated on the far left lower terminal (station port 1). The negative of the power supply is terminated on the next screw to the right of the positive (PS: + –). The

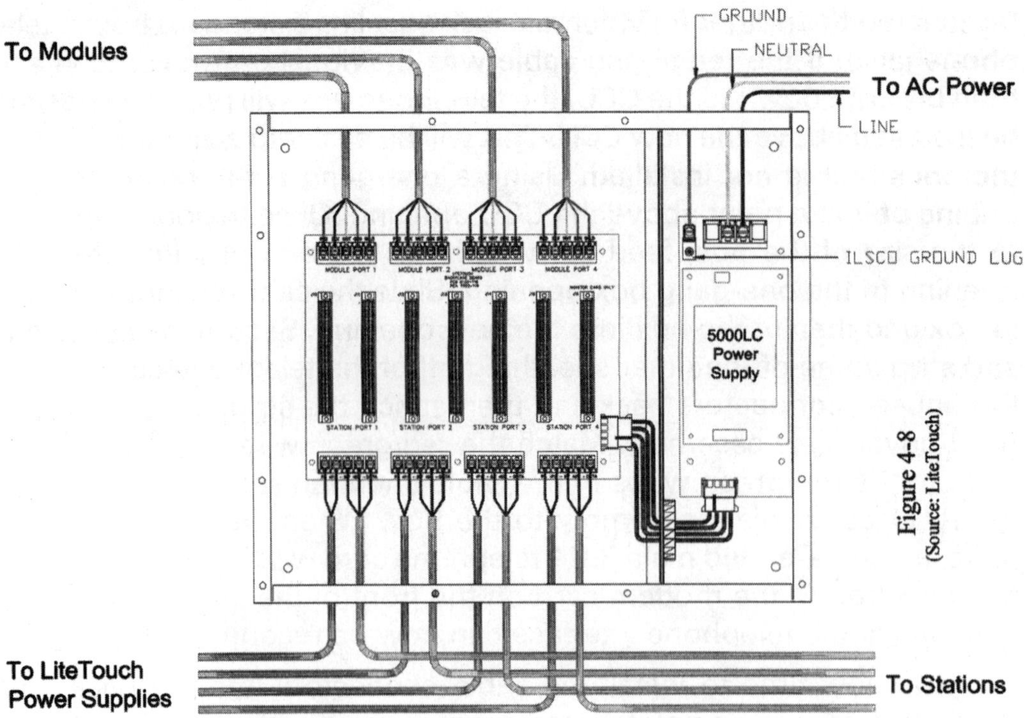


*Figure 4-8*

two switch station cables need to be connected in parallel just as they were before with a pigtail coming out of each separate colored connection (the white wire is still not used). The red pigtail wire (pos.) of the switch station is then attached to the right of the power supply negative terminal. The black pigtail wire (neg.) of the switch station is attached to the terminal to the right of positive switch terminal. The last terminal on the back plane board is for the data wire. Attach the blue pigtail from the switch station cable to the terminal marked data (SW: + – data).

The communication cable will be connected next. The red wire will be terminated on the upper far-left terminal of module port 1 (the colors are marked on the back plane board). The black wire of the communications cable will be terminated to the right of the red wire. The white wire will be terminated to the right of the black wire. The green wire will then be terminated to the right of the white wire. The last termination on module port 1 is the bare shield wire. When you have completed both the switch station and module communication terminations, there is only one cable left to run. The wiring harness that arrived with the 5000LC hardware package will now be installed. This harness has polarized connectors at either end so it can only be inserted into the boards in one direction. Plug one end of this harness into the back plane board on the left side of the enclosure. Now, run the other end of the harness through the grommet located in the partition and then plug the connector into the power supply board at the bottom. The wiring to the back plane board is now complete.

At this point, ask the electrician to connect the high-voltage 12-2 to the 20-amp circuit inside the electrical panel. You can also turn on the three control module breakers at this time. The next step will be to measure the low-voltage power at the switch station and power supply terminals. It should now be noted that the 5000LC CCU does not require an input from the single power supply located in the control module enclosure. The CCU receives its operating power from the power supply located in the right partition of the back box. The single power supply cable located inside the back box will now only power the switch stations.

While measuring the dc voltages on the left side of the partition, make sure that you do not come in contact with the high-voltage terminals located on the right side of the partition. While the lid is removed, make sure that no one is able to access the high-voltage right-hand side.

Since damage can occur to system cards from problems related to the switch station bus, it is a good time check this area before the cards are installed. With your dc voltmeter, place the red probe on (PS +) and the black probe on (PS –). Do this measurement on the wired switch station port 1. The voltage should measure between 15 and 24 volts, depending on how the system is loaded. In this particular installation, the load should be rather light, so a reading of around 18 volts or higher should be expected. Next, move the red probe to (SW +) and the black probe to (SW –). This voltage measurement should not be that different from the (PS +) and (PS –) readings. The maximum acceptable difference with a large switch load is around 1 volt or a little less. With no load, this reading should be around one-half volt. Now, leave the black negative probe where it is (PS –) and place the red positive probe on the data terminal. This measurement should not be more than 1 volt less than the (SW +) and (SW –) measurement. If this reading is more than a volt less, something is wrong on the switch station line. It could be a switch or a station wire problem. An example of a data line problem would be a (SW +, SW –) measurement of 18 volts and a (Data, SW –) voltage of 15 volts. A data voltage of 3 volts less than the switch voltage signifies unhealthy communications between the switch and CCU. We will cover more scenarios and repairs in the troubleshooting section. For now, we will consider the voltages to be acceptable on station port 1.

The power supply board should be tested next. You will notice a power LED marked D1 on the power supply board. This should be lit when the circuit breaker is turned on. If the breaker is on and this LED is not lit, turn the breaker off and check the two board fuses for continuity. If the LED is lit, locate the four-pin connector J1. The two inner pins are (+5) volts dc and the outer pins are ground. Make sure these dc voltages are present at the correct pins. Parts of this board contain high voltage so use care when working in this area. When all of the terminations have been checked for shorts and correct voltages, the lid can be screwed onto the front of the back box. You should be able to look through to the back plane board and see the card connectors and the slides that guide the cards to them. Verify that the switch station and control module cables are not obstructing this card slot area. The next section that will be addressed is the installation of the system cards. At this point, turn off all of the four circuit breakers that power the entire automation system.

## Installing the Cards

We will be installing four cards into the back plane board. The first one will be the monitor card because this is the only one that is required to fit into an exact spot on the back plane board. With the power turned off, remove the monitor card from its protective covering. Position it so that the power switch is located near the bottom. Gently guide the top and bottom of the card into the rails of the right-most slot position. Slide it to the rear and verify that the card will mate easily onto the back plane connector. Push gently on the card until the face meets the lid, then push the securing latches in at the top and the bottom. Verify that the red power switch is in the off position. The monitor card is now installed. The rest of the cards can be installed in any order, but LiteTouch prefers a standard order to help with future technical support. We will now install the next card in the standard order prescribed.

Remove the C2000 card from its protective package and then verify that the power is off to the 5000LC CCU. Line up the top and bottom (top is where the status indicator is) with the rails next to the monitor card. Gently slide it to the rear and make sure that the back plane and card connectors line up correctly. Push gently until the faceplate meets the lid. Secure the top and bottom latches and the C2000 card is then installed. So far, the monitor card is placed farthest to the right on the face of the controller and the C2000 card is installed right next to it. The right-most slot is configured in such a way that if any card other than the monitor card is placed inside, it will be rejected. If extreme force is used, other cards will fit into the right-most slot, but the system will not be configured correctly if this were to happen.

With the power still off, the CPU card will be installed next. Align the card with the rails that are next to the C2000 board. Make sure the status indicator is located near the top and then carefully slide the card to the rear back plane board. Align the connectors and push gently on the card until the faceplate meets the lid. Secure the top and bottom latches.

The last card to install is the modem. As with all of these cards, they are electrostatic discharge (ESD) sensitive. The modem card is especially sensitive to this effect and is shipped in a special ESD-protective package. Carefully remove it from the package and align it on the rails next to the CPU card. Slide it to the rear back plane and verify the connector alignment. Push gently until the faceplate meets the lid. Secure the top and bottom latches. Now, attach a double-ended male RJ11 cord from the telephone jack you installed earlier, to one of the RJ11 female jacks located on the face of the modem card. All of the system cards are now installed. The remaining expansion slots should be checked to make sure that their blank covers are securely latched. The next step will be to apply power to the system and then go over the functions of each card and their individual LED activity indicators.

For this part of the 5000LC install, we will consider that the switch/load and job programming have been previously loaded into the memory components at the factory. In the next section, we will be programming the 5000LC from the LiteWare design tool and program task, so this preprogrammed CCU will only apply to this section.

Turn on the four circuit breakers that power the entire 5000LC system. Then, turn on the red power switch at the 5000LC controller. This is located on the monitor card. The power LED on the monitor card should illuminate and LED feedback from each card should be apparent. The four LEDs located on the monitor card above the power switch and the eight LEDs located near the top of the card should start cycling. This LED activity indicates certain startup tests are running. The connection and CPU cards should have scrolling figure eight LED patterns being displayed at this time, which indicates that the cards are communicating with each other. If they are not displaying this figure eight, but instead a letter C, F, L, or some other letter, the cards are not operating correctly or are in a special mode. If a C is flashing, it shows that it is not able to communicate with another card. If the F is flashing, the FIFO, or buffers, are full. If a letter L is seen flashing, the card is in Lode Mode, which means it is waiting for a code update. More about dealing with these lettered codes will be discussed in the troubleshooting section. For this part of the 5000LC install, we will consider that no problems have arisen.

We will now discuss the monitor card and its LED indicators. The four LEDs above the red power switch represent the four switch station ports. The top LED is for station port 1, the second indicates station port 2, the third is port 3, and the fourth indicates port 4. Since, on this installation, we have only used port 1, this will be the only LED showing any activity during normal operation. If a switch button is depressed somewhere in the house, a series of two brief flashes will be emitted from the top LED. The first flash will show the information coming from the switch station button, the second flash will indicate LED update information being sent back to the switch station. If the single power supply in the control module panel is not energized or working, the top LED will stay lit permanently. The port LED will also remain lit if a switch's data line is being brought low (bad switch). This LED can be a good troubleshooting tool for switch station power problems.

The eight LEDs located near the top of the monitor card are used to indicate control module communication. During the initial startup, the C2000 card will cause the four LEDs on the left to flash while it searches for control modules. When the modules are found, the four LEDs on the right will respond by flashing the corresponding port LED that they are located on. In this installation, port 1 would be the only response flash seen. The four LEDs on the left indicate messages being transmitted to the control modules. The four LEDs on the right indicate messages being received from the control modules. The top left LED will light when control module port 1 is transmitted to and the top right LED would light when a response is received from a control module on port 1. The monitor card should always be replaced with both the power turned off and the red power button on the card in the off position. Monitor card troubleshooting techniques will be covered in the troubleshooting section.

The C2000 system card acts as an interface between the CPU card and the 2000 control modules and switch stations. It also is used as a backup to the CPU in case it fails. The C2000 has two items located on the card face—a seven-segment status display and a serial-port connector. The seven-segment display indicates a snake-like pattern (or a figure eight) when the card and system are operating normally. If a problem or a special situation arises, this display changes to indicate a flashing letter C at the rate of 1.5 times per second, which means that the CPU card is not responding. It can also indicate the flashing letter L, which means the C2000 is ready to upload new firmware. Displaying the flashing letter F means that the C2000 transmit FIFO, or buffer, is full. If a flashing letter D is displayed, it means that the C2000 is in an invalid mode (not load or operate). The second item located on the face of the C2000 card is a nine-pin serial-port connector. The nine-pin serial-port connector can operate with any straight-through (no pin swaps) nine-pin serial-port connector and cable assembly. The only pins that are used in this configuration are 2, 3, and 5. This serial port is now used to upgrade the system firmware. This card should always be replaced with the power to the CCU turned off. C2000 card troubleshooting techniques will be covered in the troubleshooting section.

The CPU card has three items on its face. It has a seven-segment display like that of the C2000 card. However, it has two nine-pin serial ports. These two serial ports do not have any differences between them. You can use the top or bottom port for job programming, real-time control, and upgrading the firmware. One of these ports on the CPU card will be used to upload the LiteWare job program in the next section. Like the C2000, its seven-segment display indicates a figure eight or a snake-like pattern when it is operating normally. If there is a problem or a special situation, the following letters will be displayed:

C indicates that the C2000 card is not responding.

D means that the CPU card is in an invalid mode.

E indicates the receive FIFO or buffer in the modem card is full.

F means the receive FIFO or buffer in the C2000 card is full.

J indicates that the job programming has invalid CRC.

L means that the CPU card is ready to upload new firmware.

The CPU card should always be replaced with the power turned off. CPU card troubleshooting techniques will be covered later in the troubleshooting section.

The modem card has four items on the front card plate. It has the seven-segment display like that of the CPU and the C2000 cards. In addition, it has a nine-pin serial-port connector that is used to upload the voice menu, upgrade the firmware, and carry out real-time control. It uses a straight-through serial-port cable/connector and only uses pins 2, 3, and 5 to accomplish the data uploading. The modem card also has two RJ11 jacks. Although the RJ11 jacks are identical, they are labeled differently to avoid confusion. The seven-segment display indicates a figure eight or a snake-like pattern when the card is in normal operation. When there is a problem or special situation, a flashing letter is displayed 1.5 times per second. A flashing letter C indicates that the CPU card is not responding; D means the modem card is in an invalid mode; F indicates that the modem card transmit buffer is full; and L means that the modem card is ready to upload new firmware.

The next area deals with a specific instruction sheet that LiteTouch has issued regarding the installation, setup, and testing of the modem card.

## *Installing the 5000LC Modem Card*

(Source: LiteTouch)

Warning: The 5000LC Modem Card is shipped in an ESD-protected package. Please observe all precautions to prevent accidental electrostatic discharge while handling the card.

**Do Hardware Installation**

1) Turn off the power to the 5000LC. (Be sure to inform everyone concerned that the system will be inoperative for a few minutes.)
2) Remove one of the 5000LC blank panels. (The Modem Card can be installed into any vacant slot.)
3) Remove the Modem Card from its box and EDS-protected wrapping.
4) Slide the Modem Card along the plastic rails. (Same procedure as any other 5000LC Card.)
5) Plug it into the Back Plane Board and snap in the nylatches to secure it.
6) Attach the phone line to either of the two RJ11 phone jacks.
7) Turn on the power to the 5000LC.

**Verify Proper Voice Mode Operation**

8) Verify that each of the CPU, C2000, and Modem Cards is snaking in the figure 8 pattern.
9) The modem card comes programmed with the following factory defaults. (Currently, the user cannot change these defaults.)
   a) Number of Rings: 2
   b) Voice Mode Password: (contact LiteTouch for this number)
   c) Data Mode Password: {see LiteTouch}(Current version of firmware does not use this password.)
   d) Voice Mode Timeout: 60 Seconds
   e) Data Mode Timeout: 180 Seconds
10) Dial the Modem Card with a handset (regular telephone).
11) It should pick up the line after 2 rings and respond with "Enter your security code."
12) Press "(contact LiteTouch for this number)" on the handset.
13) The Modem Card should respond with "Enter the station ID."

14) Enter the address of a known station (38, for example).
15) The response should be "Enter the button number" if the station is programmed, or "Invalid Station" if not.
16) Enter the button number (1, for example).
17) The response should be "Off" or "On" if the button is programmed, or "Invalid button" if not.
18) If the button is valid, enter the same button number again.
19) The response should be "Off" or "On" and the action should be the same as if the physical button had been pressed.
20) Terminate the connection by hanging up the handset.

**Verify Proper Data Mode Operation**

21) In order to verify proper data mode operation, a modem must be properly installed in the PC. Proper installation can be verified by running any Windows Telephony Program (such as Microsoft's Internet Explorer). Please verify that the PC modem works with other software packages prior to contacting LiteTouch for Technical Support!!
22) Run CP5000.
23) Select "Communications...Dial Modem" to bring up the CP5000 Dialer.
24) Enter the phone number of the Modem Card in the field labeled "Phone No."
25) The number that will be dialed is indicated in the field labeled "Number that will be called:".
26) Select the correct modem and address (if there is more than one available) in the fields under the "Line to Use" group.
27) Press the "Dial" Button and dial up the Modem Card.
28) The PC modem should dial up the Modem Card. After two rings the Modem Card will pick up the line and respond with "Enter your security code." After approximately 8 seconds of no sound, the modem card will send a beep and begin negotiating a data connection with the PC modem.
29) Once the modems have properly connected, the dialer will disappear.

30) Select the "Communications" main menu option and verify that the first submenu option is "Hang up Modem." If this is the case, then the two modems are connected.
31) If the modems did not connect, then something went wrong with the connection. Check for any error messages in the dialer message box by selecting "Communications...Dial Modem" again.
32) Most Windows PC Modems will display a modem connection icon in the system tray of the Start Menu Bar (usually located in the lower right-hand corner of the screen). Double-click on this icon and verify a reasonable connection speed up to 33k.
33) Press the "Get" button to get the 5000LC date and time. If an error message is displayed, verify that all steps in Part 1 were properly completed.
34) Next, select the "Control Station" tab.
35) Enter a valid station address in the "Type Station Address" field, and press "<Enter>".
36) The LEDs above the buttons should match the physical station LEDs.
37) Press a button that is programmed and verify that it operates just like the physical station. Note: Because of processing speed through the modem, the response will be slower than a normal button press.

**Verify Proper Data Download**

38) Press the "From 5000LC..." button to test downloading the job programming.
39) A "File Save As" dialog box will be displayed. Enter a filename (garbage.prg for example).
40) Press the "Save" Button.
41) A progress bar will indicate download progress. Upon finishing, the file will be saved to disk.
42) This downloaded file can be compared with a known good file using the FC (file compare) DOS utility. Assuming the known good file was named good.prg and the downloaded file was named garbage.prg, change to the subdirectory containing both files and type "fc /b garbage.prg good.prg "<Enter>".

43) If the response is "no differences encountered", then the download process completed successfully.

If everything works properly after following the steps listed, the modem card is properly installed. (Source: LiteTouch)

This completes the section on installing the 5000LC hardware system. The next section will address the programming part of this system installation by using the LiteWare design tool and program task. More information regarding the 5000LC will also be covered later in the troubleshooting chapter.

You probably have noticed from performing the standard 2000 and 5000LC hardware installations, that these systems have much in common. This commonality allows you to transfer the experience and knowledge from one system to the other. The next section on programming is another example of where your previous experience with the Standard 2000 CCU program environment will ease the transition into the LiteWare 5000LC environment.

## LiteWare Programming

LiteWare is a 32-bit, Windows® 95/98/NT program. This program requires a 486 or better processor and at least a 66MHz clock. It needs about 4MB of your hard drive to operate. The LiteWare software includes two distinctly different internal programs. The program task actually creates the lighting control program file that will be uploaded to the 5000LC CCU. The design tool software is used to create the load and switch assignments that will later be imported to the program task for uploading. This software can be used to design and upload switch/load information for the 5000LC CCU. In addition, it can also design and export switch and load information for the Standard (2000) CCU. The LiteWare programming environment is a little different from the Standard CCU programming environment that we covered earlier. Since it is Windows-based, it gives an enhanced graphical view of the switches and loads that you are processing. Additionally, it has many features not included in the previous versions of LiteTouch programming software. To obtain this software you can visit the Web site LiteTouch.com and download the files containing the LiteWare program, or contact LiteTouch directly for the program CD. Detailed instructions accompany both the download and the CD versions. The following are instructions for installing LiteWare onto your computer using the CD version.

First, close all of the Windows applications before starting the installation. Insert the LiteTouch CD into your CD-ROM drive. In some instances, the install will start automatically. If it does not, click on the Start tab and select Run. Type D:CDRUN (D is typically the letter for CD-ROM drives, but if you have a different letter that denotes the drive, use that) and select OK. Click LiteWare Install from the window that appears. Another window will appear after the LiteWare Install selection; click Next. The next window will inform you where the file will be copied. Click Next if you accept the default. A window will then appear that says that the install is complete. It will give you two choices to either restart your computer now or to wait until later. If you wait until later, you will not be able to access the LiteWare program until you do a restart. Click Finish to complete the install. Since the LiteWare program is graphically orientated, it may be helpful to read this part of the book while you view the software.

As you enter the LiteWare environment, you will be prompted by a dialog box to input a new drawer and job name. This will only happen when the software has just been installed; however, you can input new drawers and job names at any time while you are inside the LiteWare environment. Type a drawer and job name and select OK. The next dialog box that will appear is Job Preferences. The information in this box will set the initial parameters for the programming inputs. Select 5000LC from the CCU Type field, since this will be our programming source CCU. The next field will give the option of 110 volts or 220 volts. Select 110 and leave the default setting of 1920 watts. The next field is Control Stations; select Actual from the two options listed. The Estimated Item or Part Number field can be left as it is at this time. Select OK, since all of the pertinent information has been entered into this dialog box. Another dialog box will appear that asks you to enter the name of the enclosure schedule. Type Utility Room and then select OK. The load-input screen will appear next.

### Inputting the Loads

The load-input screen will allow you to enter all of the loads that will be assigned to the future switch stations. The first field on this screen says Load# and this will be your first input. The load information that has already been compiled will be used again (Figure 4-9). After the Load#, type 101 and then mouse click over to the Load Type field (we will not use the Fixture field, so skip past this). This is a drop-down list and will include many different selections. Select Incandescent from this list and then mouse click over to the Watts input field. Type 100, then click over to the QTY field and type 1. There is only one outside light on this house, but if there were more, you would input how many lights were connected to this load circuit. Click down to the Function field and use the drop-down list to select Switching. The next drop-down list field is Plan. This field is for houses that have more than one floor plan. Since the print we are using has only one, we will pass this field and go on to the Area field. The Area field is also a drop-down type. It retains the last entry in memory so you can recall it for the next load in the area. This house only has one load per area, so this feature will not help us much in this installation. Type Outside in the Area field and click over to the Description field. Type Outside Wall Light into this field. This load is now complete, so click on Add at the right corner of the load screen. The information will then be transferred to the load area below.

The next load input is 102, so type this into the Load# field. Click over to the Load Type drop-down list and scroll until you find Switched Outlet. Select Switched Outlet and then click over to the Watts field. Type in 200, then click over to the QTY field and type 1. Click to the Function field and select Switching. Skip the Plan field and click to the Area field. Type Bedroom #1 and then click over to Description. Type Bed #1 Outlet and then click Add. This load information should be transferred to the second load area below.

Continuing on to the next load in Figure 4-9, we will input 103 in the Load# field. The Load Type drop-down list should include fluorescent lights, so select that and then click over to Watts. Type 200 and then type 1 in the QTY field. The Function field is next, so select switching and click over to the Area field. Type Kitchen there and then click

## Control Module Enclosure Schedule

Module address: 01 Module type: Relay 6-channel No.of enclosures: 1

| MODULE POSITION | LOAD # | LOAD LOCATION | WATTAGE | CIRCUIT SUPPLY |
|---|---|---|---|---|
| 1 | 101 | Outside Wall Light | 100 | 1- 20 amp 110v circuit |
| 2 | 102 | Bed #1 Outlet | 200 | |
| 3 | 103 | Kitchen Light | 200 | |
| 4 | 104 | Bed #2 Outlet | 200 | |
| 5 | 105 | Den Outlet | 200 | |
| 6 | 106 | Master Bed Outlet | 200 | |
| | | Total Load: | 1100 | |

Module address: 02 Module type: Dimmer module

| POSITION | LOAD# | LOAD LOCATION | WATTAGE | CIRCUIT SUPPLY |
|---|---|---|---|---|
| 1 | 201 | Foyer Light | 200 | 1- 20 amp 110v circuit |
| 2 | 202 | Dining Light | 200 | |
| 3 | 203 | Bathroom Light | 200 | |
| 4 | 204 | Hallway Light | 100 | |
| | | Total Load: | 700 | |

| | |
|---|---|
| POSITION 3<br><br>POWER SUPPLY MODULE (SINGLE)<br>OUTPUT: dc | 1-20 AMP<br>110V CIRCUIT |
| POSITION 4<br><br>DATA INPUT MODULE (MOMENTARY)<br>CONTROL: Security Interface<br>Switch Station Address: FF-1 | N/A |

*Figure 4-9*

to the Description field next. Type Kitchen Light in this field and then click Add. You should now see three loads and the corresponding information for each in the list below. You probably have noticed that the fields will leave the information from the last load in them as you are starting your next load. This feature can save some time if the next load shares some of the same information. Continue to input the information for loads 104, 105, and 106. These will all be switching loads because they are controlled by a relay. Loads 201, 202, 203, and 204 will all be dimming loads and the information from these can also be entered into the fields of the load screen. When these are all entered in the load list, click on the File drop-down list at the top of the screen and select Save Location. The load information for the house is now complete and will be transferred automatically to the Switch Design section.

The next section will cover the programming of the switch control stations. You may wish to reference back to the switch station design guides used in the earlier Standard 2000 install.

### *Inputting the Switches*

Click on the View drop-down list next and select Station Design. A dialog box will then appear and display "No stations have been entered for this job." Click on OK and a New Station dialog box will appear. Type 1 (the print number for the Foyer switch), then click on Description. Type Foyer and then click down to the Number of Gangs field. Make sure that 1 is already selected, then click down to Buttons Per Gang. The Foyer is designed to have a five-button switch, so use the drop-down list and select 5. The last drop-down list is Address and it should have the 01 already in the field area, so skip that field and click OK. This will bring you to the switch programming page.

On the switch programming page, you will see a five-button configuration to your right and all the loads that were just programmed to the left. Above the loads is the input area. You will notice that the first position on the switch configuration has an LED that is lit. If you click on any one of the five buttons, the lit LED will transfer to the one selected. This is how each button will be identified while you enter information into the program. Make sure that the LED is lit on position one before you start inputting data. The order of the inputs is not that important, but for speed, you may try to develop a certain order to the progression. Let's start by assigning a load to switch 01-1 (refer back to Figure 1-3 to help with the programming). First, click on load 101 and then click on the Assign tab. You will see that load transferred above to a box. The load is now assigned. If a mistake is ever made regarding the load, just highlight the load that you do not want assigned and click on the Unassign tab. The next step will be to program a function for the button. Click on the Function drop-down list and select Toggle On. You will notice that there is not a separate listing for Toggle like that of the 2000 software we previously programmed. In this situation, Toggle On will act just like a regular toggle switch. It will turn the load on/off and will also show the correct LED status. This function switch will also act like a Toggle master by controlling many loads in the same way. After you have selected Toggle On, click on the Engraving field. Type Outside on the first line, click to line below and type Light. This button is complete, so click button number 3, the next one over.

With the number 3 LED lit, click on load 201 and then click on the Assign tab. Verify the load transfer and then click on the drop-down Function list. Select Dimmer from the list and then click on the first line of the Engraving field. Type Foyer and this button will be complete.

Click on the number 5 button and verify that the LED is lit. Click on loads 103, 201, and 204. Add them to this button by clicking the Assign tab. Click the Function drop-down list and select Scene. Then, click on the first line of the Engraving field and type Kitchen. Click down one line and type Path. Now, click on load 103 in the assigned load box and it will highlight. Next, click in the Set Level field and type 100 (relays can only be set to 100%). Click the next load down in the assigned load box (this should be load 201) so that it highlights. Click to the Set Level field again and type 75. Click back to the last load in the assigned box (204) and then click back to the Set Level field. Type 75. You will notice that the loads in the assigned box now have their corresponding level settings placed at the end of the load description line. This is a good way to verify that your level entries have been programmed.

Click on button number 7, then click on the All Loads tab. The entire lower load box should be highlighted now. Then, click on the Assign tab and all of the loads should move into the upper assigned load box. Click the Function drop-down box and select Toggle Off. Click on the Engraving field first line and type House. Click down to the second line and type Off. The function Toggle Off will turn all of the loads in the house off and then its LED will go dark. (This LED turns off when it accomplishes its function.) If any load in the house is then turned on, this LED will light. This switch will also turn on all of the lights in the house once all of the loads are off. This sounds like double-talk, but if you think about the way the function program works it makes sense. It has its main function that makes all of the lights turn off in the house. Once this function is fulfilled and the loads are off, it can then toggle them back on. Most people will just use the next button on the switch, which is House On.

Click on the last button and then click the All Loads tab. Then, click the assign tab and click to the Function drop-down list. Select Toggle On and then click to the first line of the Engraving field. Type House, then click down a line and type On.

You may notice that the engraving schedule has automatically been transferred to all of the buttons. This is a good time to check that this information is correct (Figure 4-10). In the Colors section, you can choose the button, text, and LED colors. We are using the defaults on these switches, so there is no need to change anything in this section. Click OK to exit. The next tab over is Gang and we have already made the choice regarding the number. The tab to the right of that is Style. We are using S-Series (sculpted) switches with a white prime coat for this installation, so the defaults will work fine here also. Click OK to exit. We will want to save this completed switch station, so click on the File drop-down list at the top of the screen and select Save Station.

To program another switch station (Figure 1-4), click on the File drop-down list and select New Station. The New Station dialog box will appear:

- Station Plan Number__________(type 2, then click down to the next field).
- Description________________ (type Bedroom #1, then click down to the next field).
- Number of gangs_____________(leave at default setting which is 1. Click down to the next field).

**LiteTouch Station Design Schedule**

Job: Book
Date: May 05, 2000 (11:42)

Station Location: Foyer
Drawing Number: S1

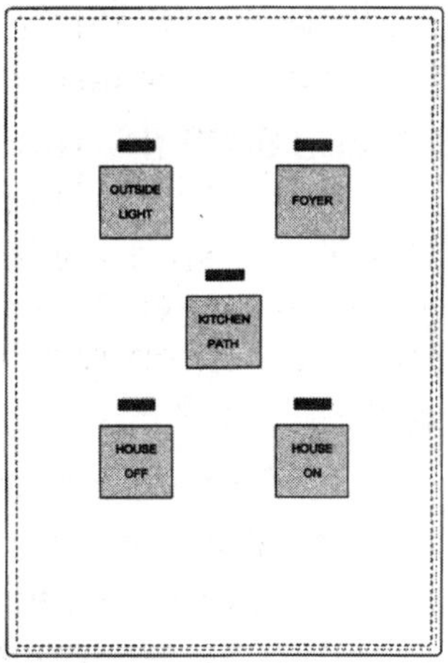


| Pos # | Button Color | LED Color | Text Color | Engraving Text |
|---|---|---|---|---|
| 1 | White | Green | Black | OUTSIDE<br>LIGHT |
| 3 | White | Green | Black | FOYER |
| 5 | White | Green | Black | KITCHEN<br>PATH |
| 7 | White | Green | Black | HOUSE<br>OFF |
| 9 | White | Green | Black | HOUSE<br>ON |

*Figure 4-10*

| Drawing # | Location | Series | Style | Colors | Gang | Address |
|---|---|---|---|---|---|---|
| S1 | Foyer | S | Sculpted | FACEPLATE: White Primecoat | | 01 |

- Buttons per gang_____________(scroll through drop-down list and select 3. Click down to next field).
- Address__________________(should be at default of 02, which is correct).

Click OK and the new switch-station programming screen will appear. The first button on the switch to program is 2. Verify that the number 2 position LED is lit and proceed to the lower load box. Select load 102 and assign it to the button. Click over to Function and select Toggle On. Then, click to the first line of the Engraving field and type Room. Next, click on the button position 5. With the LED lit, click on load 103 and then assign it to the switch button. Click to the Function field and select Toggle On from the list. Click to the first line of the Engraving field and then type Kitchen. Now there is only one more button to program on this switch.

Select the number 8 button, then click over to the load area and select loads 203 and 204. Assign them to the switch. Click to the Function field and select Scene from the list. Click to the Engraving field and type Bath on the first line. Click down a line and type Path. Make sure all of the buttons are engraved properly and that the loads and functions are correct. To save this switch program, click on the File drop-down list and select Save Station.

Continue to program switches 03, 04, 05, 06, 07, and 08 in the same manner. The Data Input module will be programmed later in the program task when the switches are imported. This concludes the switch programming until the design is imported to the program task.

### *Setting up the Control Module Enclosure*

The control module enclosure will be covered next. From the main LiteWare screen, choose the Edit drop-down list. From this list select Assignment. A dialog box will appear asking if you want to auto-assign. If you desire this, click yes. Clicking yes will let the computer assign all of the loads to specific modules using an algorithm. In a new installation where you have not yet planned any of the loads or modules, this selection can be very helpful. In this installation, we already know from our standard install what modules and load addresses we will want, so click No. At this time, click on the Edit drop-down list again and choose Add Enclosure. A New Enclosure dialog box will appear and a default enclosure name (the letter "A") will be highlighted. You can click OK, unless you want another name assigned. Now an empty four-module enclosure will appear labeled Enclosure A. Click on Edit again and select Modules from the list. An Edit Modules dialog box will appear, so click on the New tab in this box. A New Module dialog box will overlay across the Edit Modules box. In this New Module dialog box, scroll down through the modules until you arrive at an eight-channel relay. Double-check that the load address is 01 and that the enclosure position number is 1. When all of these items check out, select the eight-channel relay and click OK. You will then see the eight-channel relay in the lower unassigned box. Click the assign tab to the right and it will be transported to the assigned area above. Now, click on the New tab again and the New Module dialog box will reappear. Scroll down through the list of modules again until you arrive at the eight-channel Dimmer module. Make sure the module address is 02 and that the enclosure position is 2. Then select the eight-channel Dimmer module and click on OK. As-

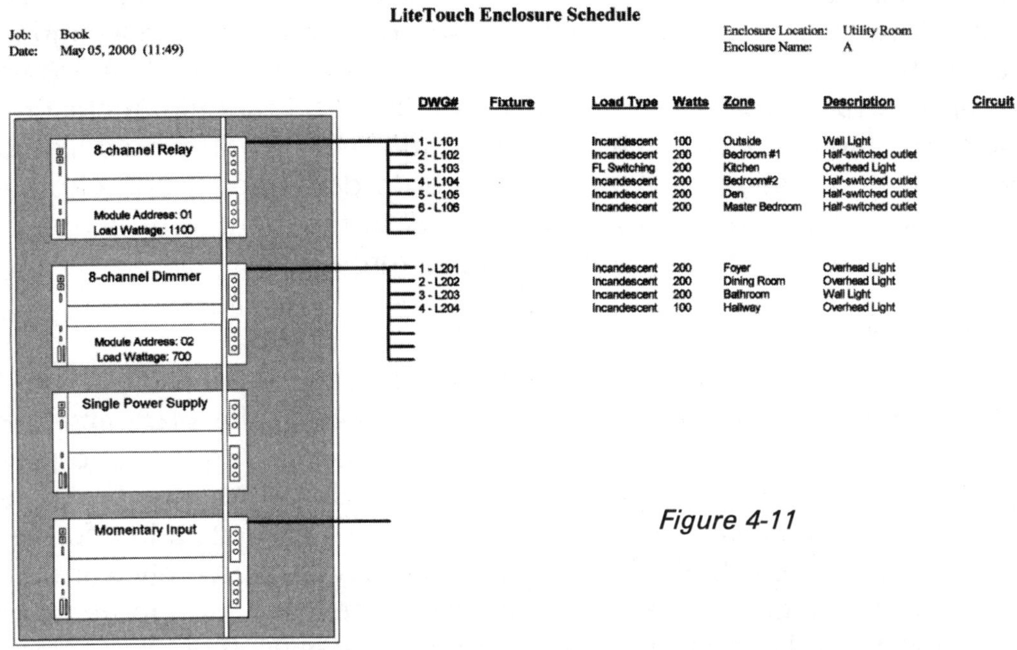

**LiteTouch Enclosure Schedule**

Job: Book
Date: May 05, 2000 (11:49)

Enclosure Location: Utility Room
Enclosure Name: A



| DWG# | Fixture | Load Type | Watts | Zone | Description | Circuit |
|---|---|---|---|---|---|---|
| 1 - L101 | | Incandescent | 100 | Outside | Wall Light | |
| 2 - L102 | | Incandescent | 200 | Bedroom #1 | Half-switched outlet | |
| 3 - L103 | | FL Switching | 200 | Kitchen | Overhead Light | |
| 4 - L104 | | Incandescent | 200 | Bedroom#2 | Half-switched outlet | |
| 5 - L105 | | Incandescent | 200 | Den | Half-switched outlet | |
| 6 - L106 | | Incandescent | 200 | Master Bedroom | Half-switched outlet | |
| 1 - L201 | | Incandescent | 200 | Foyer | Overhead Light | |
| 2 - L202 | | Incandescent | 200 | Dining Room | Overhead Light | |
| 3 - L203 | | Incandescent | 200 | Bathroom | Wall Light | |
| 4 - L204 | | Incandescent | 100 | Hallway | Overhead Light | |

*Figure 4-11*

sign this module in the Edit Module dialog box and click New again. Select a Single Power Supply from the list, check that the position number is 3, and then click OK. Assign it at the Edit Module box and click New once more. Scroll down through the list until you find a Momentary Input module, then select it. The only position left in this enclosure is 4, so this should be the number displayed. Click OK. Assign the Momentary Input in the Edit module dialog box. If you close this box, you should see a graphical version of what the real enclosure will look like with the modules installed (Figure 4-11). The relay should be on top, the dimmer second, followed by the power supply, and finally the Momentary Input. The loads will now need to be assigned to the modules.

Click on the Edit drop-down list and select Module Loads. The Edit Module Loads dialog box will appear with all of the house loads listed at the bottom. Click on the first load, 101 and then click on the Assign tab. You will see load 101 transport to the relay module position one area above. Repeat assigning loads 102, 103, 104, 105, and finally 106 to the relay module. When this is complete, scroll through the Module field drop-down list until you arrive at the eight-channel dimmer module and then select it. Assign loads 201, 202, 203, and 204 from the unassigned box below. Once they are transported in order to the above assigned box, you can close this dialog box. Then, click on the File drop-down list and select Save Assignment from the list. The module enclosure, modules, and loads are now assigned and saved. If you would like to print this information, click on the File drop-down again and choose Print from the list. You can print almost all of the input information in LiteWare. The print options will guide you through this.

### *LiteWare Terms*

Now that the switch and load information has been programmed into the design tool software, it is time to work in the program task environment. Before importing the current working file into the program task, it may be helpful to go over some of the terms used first. The program file contains all of the load and switch station data in a format that is compatible with the 5000LC CCU. It is saved separately from the design information and has no direct link to the loads, assignments or switch stations. Load group is another term that is used quite regularly in the LiteWare program task. A load group is a group of loads and their set parameters used together in a switch or timer assignment. A load group can be made up of one or more loads. In addition, parameters can be programmed into the group such as dimmer, ramp, and fade levels as well as load preset or minimum turn-on levels. Advanced User is a term that refers to one of two operation modes. Normal User is the other operation mode. When the Advanced User mode is selected, it allows the programmer to use the advanced features in the programming environment. For example, the address ID number column is added to the loads box. The Normal User selection would not let you see this particular information. Advanced User is the default setting in the LiteWare environment, so if you prefer this feature, there is no need to make a change in the program task. Throughout the book, reference has been made to the terms switch and button. Usually, these terms will be interchangeable. When a complete switch is being referred to, generally the term switch station will be used. The last term that will be explained is System Event. A system event is considered an action defined by its loads and parameters. This occurs when one of the following triggers is activated:

1) Pressing a switch
2) Holding a switch
3) Releasing a switch
4) Matching a timer's start time
5) Entering a combination (user event)
6) Activating a startup event by starting or restarting the 5000LC

Other terms will be used in the program task, but many of them have already been explained earlier in this section.

### *Importing and Working in the Program Task Environment*

Importing the file that contains the switch and load information to the program task will now begin. From the main LiteWare screen (design tool software screen), open the file that contains the programming for this installation. Once this has been opened, click on the 5000LC program tab on the right-hand part of the screen. The program task screen will then be displayed. On this screen, click on the Options drop-down list and select Import Design. The switch and load data will then be transported from the design tool software to the program task. You can verify this import by inspecting the right-hand side of the program task screen and seeing if there is a graphical display of the programmed switch stations. Scroll through the switch stations to make sure the addresses 1 through 8 are present and in the proper switch configuration. Check the bottom of the program task screen for any error messages. If any loads are unassigned or improperly entered, the import will not be successful. When the screen is devoid of any error messages, you can then proceed to the load area. Check to see if all of the programmed loads are present in the load field on the left-hand side of the program task screen. When both the switch and load import data are confirmed, we can then program the last switch station in the program task.

The Data Input module station will now be programmed. First, click on the New Station tab right above the graphic station display. A Style dialog box will then appear. Leave the "S" series station as the default and then click down to the Number of Buttons field. Scroll through the numbered list and then select 8. You may be wondering why we are choosing a standard switch station instead of an actual Data Input module. The truth is, the CCU will not know the difference, or at least not enough of a difference. The standard Data Input module has eight inputs and an address exactly like a standard switch station. Continuing on with the programming of the Data Input module, click OK after you have selected the eight-button switch station. Next, a station dialog box will be displayed. Type Data Input Module in the Name field and then click down to the Address field. Scroll down to the last address, which will be FF, and select it. Make sure that the button position is listed as 1, then click to the Function field. Scroll down this list until you arrive at the On (Timed) function, and select it. Now, click on the Program tab on the lower right-hand corner of the box. The On (Timed) dialog box will appear. The first field in this box is labeled Load Groups. Scroll down this list until you arrive at House On and select it. This choice will include all of the loads in the house, so it is a good load group to choose for this security application. Now, click on the minutes area of the Active Time field. Type 15 in this area and click over to the seconds area and change it to read zero zero. Click on Close of this dialog box and OK of the next dialog box. Check the switch graphic display area on the right of the program task screen to verify that the Data Input station is now present. Click on the number 1 position of this switch and verify that all of the loads on the left side of the screen go to a highlight status. Then, check above to the Group field to see if House On is typed on this line. Now, click on the File drop-down list and select Save. The program will then want you to input a filename for the save. Type in the same filename you have been using to program the loads and switches in the design tool software. Click on the Save tab and the program will be saved to your hard drive or disk.

This may be a good time to verify the load and button configuration of the graphic switch stations on the right side of the program task screen. Make sure that the switch station address corresponds to the correct button configuration. In addition, click on each button of each switch station and read the corresponding load information that appears to the left with each button click. Correct the information as needed.

The next section will cover items located in the Options drop-down list at the top of the program task screen. Click on the Options drop-down list and select Preferences from the list. A Program Preferences dialog box will be displayed. In this box, you can verify or input information such as latitude, longitude, time zone, and daylight saving time. At the bottom of this dialog box you will see the communications parameters. Correct port and baud rates can be changed or verified in this field. This information has to do with the communications interface between the PC and the 5000LC CCU. The port choice will be dictated by your personal computer equipment; the LiteTouch default baud rate is 9600. When you are finished with the information on this screen, click OK to exit and return to the program task screen.

To set the current time and date into the program, click on the Options drop-down list again. Select Set Time from the list. A Set Time dialog box will be displayed. Enter the time and date and click on the OK tab.

To select the timer function, click on the New Timer tab that is adjacent to the New Station tab. The first field of the timer screen is labeled Name. Enter a name that describes what the timer will control, such as Outside Light. Click to the Type field, which includes the headings of Daily, Week, Month, and Year. Selecting Daily will activate the specified load or loads every day at a specific time or sun event. Choosing Week will display a seven-named day calendar. You can click on the day or days that you want the particular load or loads to activate. When you click on Month, a 31-day calendar will appear. You can choose what days you want the timer to activate by clicking each day. Finally, the year includes both a 31-day calendar and a monthly display. Clicking on each month in combination with the days will set the timer to activate at the desired interval. After the appropriate selections have been made, click over to the Time field. You can enter hours and minutes along with the A.M. or P.M. designation. The astronomical clock function will permit you to select sunrise or sunset. In addition, you can program sunrise or sunset timers to activate XX amount of minutes before or after the sun event. The next timer field is labeled Function. Scroll down through this list to choose on/off, on, off, and user functions. Once you select one of these functions, a dialog box will appear. In this box, a load group will need to be chosen for the timer activation. When the load group is selected, click the Close tab to return to the Timer screen. Then, click the OK to exit to the program task screen. Check the graphic display to the right. Grouped along with the switch stations should be

**View is looking at connectors head on.**

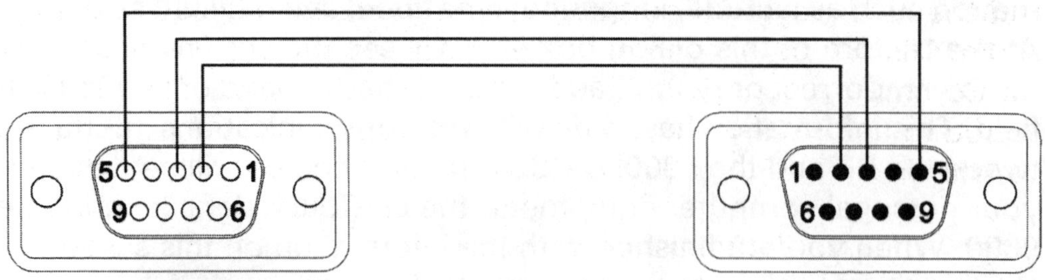


**FEMALE**

**MALE**

**COMPUTER SIDE PINOUT**

Pin 2 - Rec. Data
Pin 3 - Trans. Data
Pin 5 - Signal Ground

**LITETOUCH PINOUT**

Pin 2 - Trans. Data
Pin 3 - Rec. Data
Pin 5 - Signal Ground

**<u>Note:</u> Pins 1, 4, 6, 7, 8, and 9 are not used by the 5000 LC.**

*Figure 4-12*

the timer graphic display that you just programmed. It should be labeled with the name you entered into the Timer box. Clicking on the graphic timer icon will display the programmed parameters. At this time, it may be a good idea to click on the File drop-down list and select Save. There are many more options available in the program task and they will be covered later. For right now, however, we will keep the programming simple.

This will conclude all of the necessary inputs into the program task. It is now time to prepare to upload this program into the 5000LC CCU. The communications cable required to accomplish the upload to the 5000LC has a different wiring configuration than the cable used to upload to the Standard 2000 CCU; therefore, a new communications cable will need to be acquired. Figure 4-12 shows the cabling and pinout requirements. LiteTouch can supply this cable, or you can purchase it elsewhere. Pins 1, 4, 6, 7, 8, and 9 are not used by the 5000LC.

Once the correct serial cable has been connected to both the PC and the nine-pin serial port on the CPU card, the 5000LC upload can begin. Click on the Option drop-down list and select Export to Controller. Communications will then be established and the file will be exported to the 5000LC CCU. A graph will display the details of the data transfer from the PC to the controller.

### *More about LiteWare*

The LiteWare environment has many programming alternatives to choose from. The programming that you have just completed in LiteWare applied many of the important elements of the software. However, some of the special features of LiteWare were not implemented into this program. It is much like the software that resides on your personal computer; the full potential of it is not always used. Sometimes this is true because you simply do not have the need to scope out its true potential. In LiteWare, the more potential you uncover, the more business you will probably create for yourself. Specialty programming is not practiced by as many technicians as you might think. Becoming proficient at it may be a great advantage and give you the edge in home automation that will set your skills apart. There is not any set way to specialty program in LiteWare. The only limit set in this environment is by your own imagination. It helps to have a programming background, but it is not necessarily a requirement. Before we begin the advanced programming section, it may be helpful to go over some more details of the LiteWare environment.

Load groups were mentioned briefly during the previous programming example. More details about them will be discussed now. Load groups can be created by the programmer or can be automatically created when the initial load information is imported into the program task. The main purpose of using load groups is to reduce the time it takes to program. In a large installation, you will have many switch stations to program. Each button requires a load address assignment and more often than not, a single button will have many of these assignments. Inputting the same load addresses over and over is a waste of your valuable time, so having specific load groups to input will ease the workload. You may have noticed how quickly the Data Input module was programmed with the use of one such load group. In addition to reducing the load inputting, the load groups also reduce the programming time involved with assigning the parameters. For example, a single dimmer ramp time can be assigned to an entire load group. In the program task, there are fields that allow you to view, add, delete, or modify each load group. The Group field will show the name of the load group that has been selected. If you click on this Group drop-down list, it will display every load group that is available to modify. Using the (+ – ...) tabs to the right of the group field allow you to create a new group with the (+) tab and delete a group with the (–) tab. The (...) tab lets you add/modify the ramp, dimmer, fade, or preset/minimum turn-on levels for each load in the selected group.

You can also use the Drag and Drop capability to create another load group. From the program task environment, click on the desired loads inside the loads box. When all desired loads are highlighted, move the mouse to one of the loads, press the left button and hold it. Drag the mouse pointer over to the graphic display box on the right-hand side of the screen. Let go of the mouse button when you arrive at the switch or timer that you would like the group assigned to. If a function has not yet been assigned to the button or timer, a dialog box will appear for you to make a selection. Once you have chosen the function, click on the Program tab. This will display a Function dialog box. The new load group is then created using the same name as the button or timer that you dropped the group on. The only difference is that the name now has a numbered identifier at the end of it. For example, if a switch is labeled House Path, then the new name will be House Path_1.

Low-voltage loads (dry-contact module closures) are not defined as loads in the LiteWare environment. These loads must be added separately to the program task. At the program task screen, click on the New tab above the loads box. A dialog box will then be displayed. The first field needs to have the low-voltage named typed in it. For example, if a low-voltage contact closure controls a garage door, type Garage Door in this field. The next field down is labeled Area. Type in the area where the load is located, then click down to the Module address field. We have not covered low-voltage relays yet, but it has a control module address like any other load control module. Input the two-digit address, then click over to the Output field and select the number that corresponds to the low-voltage module output number. For example, the first output on low-voltage module 88 is connected to the garage door, so select the number 1 for this field. After this has been completed, click OK and this will add the load to the list.

You may have noticed a function named Switch Pointer while working within the LiteWare program task. This function is used when programming two or more switches. If the information for one switch needs to be duplicated from another, this switch pointer function will point the new switch to the original switch's function and load information. There is no need to program the same information into the new switch. Simply point the new switch to the existing switch. You can also copy the loads and function from one switch to another. As with most software, there are many programming techniques that accomplish the same results.

Another LiteWare term that may already be familiar to you is User. This User function, which is part of the specialty programming that was mentioned earlier in this section, is utilized to create custom instructions. A User function can be assigned to a switch or timer, and it will give extra instructions to the regular program. We will cover this function in greater detail later in this section.

A Combination function uses a security code in addition to a button press. This program stops unwanted load activations by unauthorized individuals. This code can have up to 15 numbers that are limited by the positions of the switch station. For instance, a four-button station can only use the numbers 1, 3, 7, and 9 in the security code.

The Motion Control function is primarily used for controlling forward and reverse motor loads. This is primarily a timing function and only stays activated for a specific programmed duration. This function can be used for motorized curtains and blinds. The next area of discussion will address the specialty programming that was mentioned at the beginning of this section.

### *LiteWare Advanced Programming*

This section illustrates the custom program capabilities of the LiteWare software. These custom capabilities cannot be programmed with the standard LiteWare functions. These special instructions are implemented by using certain User Programs. There are three types of

User Programs:

1) User Function (accessed from the Function field when programming a new switch)
2) User Event (accessed from the Edit drop-down list in the program task)
3) Startup Event (accessed from the Edit drop-down list in the program task)

The User Function can have custom instructions written for three separate actions: press, hold ,and release. Switch buttons and timers can be programmed with independent instructions with regard to their function action. The timer is usually programmed with regard to its start and stop action.

User Events are activated when the Combination function is applied (when an input code is used). A Combination function is the only function that can initiate a User Event program.

A Startup Event is a group of instructions initiated when the 5000LC is powered up or when it is reset. A Startup Event program is not necessary for a power-up or a reset. The controller, however, will use the last-known settings when it initially comes back on line. If any of the cards of the 5000LC were removed during a power-down, the last known settings will not be present when the unit is powered back on. When a Startup Event is programmed into the system, it will always execute its instructions whether it was powered up or reset.

Program functions define the actions for certain load groups. These are some of the commands that are used:

SetLoadsOn (sets the specified loads to On)

SetLoadsOff (turns the specified loads Off)

ToggleLoadsOn (toggles loads On)

StartRamp (used for dimming)

StopRamp (used for dimming)

You will notice that there is no space between the words of the command line. This is the way the terminology needs to be written in this custom environment. These program functions are normally mated with the load group so that a complete program line will be acted on and understood. It is very important that these program lines be accurate. In most instances, a system can act very strangely when improper program functions and commands are used. Infinite loops are a common mistake and can actually lock up the 5000LC CCU. However, careful planning and proper programming can make the automation system do some wonderful tasks. You may write a program or a Macro sometime (the macro will be discussed later) and be hesitant to upload it to the CCU for fear of what the results will be. Call LiteTouch and see what it thinks about the program in question. You never know; the company may like it so much that it will tell others about it. Using common sense and logic are your best tools when programming.

We will now address an important term that is used quite often in programming—the Stack. The Stack is where values are stored. After a program is executed, it will place the information it received on the stack. When a new value is moved to the stack, it will be placed on top of the previous value in the stack. Generally, the information located on the top of the stack will be the value you will work with most.

Querying or changing system values is mainly about receiving information and then doing something with it. The "GetLoadValue (loadID1,groupID1)" program line will read the load level with the ID of 1 and the group ID of 1 and will place this value on the stack. The "SetLoadValue(loadID2,groupID2)" program line takes that value and uses it to set the load level for the load of ID 2 in group ID 2. When you use these functions to actually make an automation system do something interesting, these command lines will take on a new meaning.

Arithmetic functions are used in addition to the program functions. Changing values in the stack can be accomplished by the use of operators, commands, and values. An example of this will be given: 7 – 2 = 5 can be entered as:

```
enter (7)
enter (2)
sub
```

In this situation, the first enter places the 7 at the top of the stack. The second enter now places the 2 at the top of the stack. The sub command subtracts the top two numbers on the stack and replaces both of them with the number 5. The following example uses the Get command to do an arithmetic program:

```
GetLoadLevel(loadID2)
enter(10)
add
SetLoadValue(loadID2,groupID2)
```

This operation reads the load level for load ID 2 and places the value on the stack. The 10 is also placed on the stack; then it adds the two values. It will then replace both values with the new value. Finally, it will change the load level for load ID 2 in group ID 2 to the new value.

A State variable is used only once in a user program. It can be set to any numeric value. For example, Setstate(5). The 5 is the state variable and when the program has ended, this value will return to zero.

A Global variable is system-wide and user-defined. All programs on the 5000LC can use the same global variable. Any program can reference it and use it once it has been set. Many global variables can be created throughout the program.

A program will be executed when one of the following activities occurs:

- A Startup Event (the system is powered on)
- A User Event (a combination code is entered)
- A User Function (a specific button is pressed, held, or released. Also, when a timer is activated)

Once a program is initiated, the 5000LC opens an interpreter to run each program. Each interpreter has its own separate stack and reads each statement line one at a time. In addition, it will end the program when it reads the statement "term." This acts much like the End statement that was used in some of the earlier programming languages like Basic.

Some of the rules regarding proper syntax will be covered next. Statements and commands need to be placed in a single line. Use one statement or command per line. In addition, as stated earlier, spaces should not be used in the program line. When using a function, always use the drop-down list to select an item. This will alleviate any entering errors. When using arithmetic operators or commands always use lowercase letters. You can add comments to a program, but make sure you use a semicolon at the beginning of the comment statement. The interpreter will ignore any typing with a semicolon ( ; ) in front of it.

Conditional statements are a very important part of the programming in the LiteWare environment. These statements can have a great practical value in home automation. For instance, say that you have a home with outside storm shutters controlled by the LiteTouch system. These shutters cover the out-swinging doors and the out-swinging windows. Sensors could be installed in such a way that when a door or a window is left open, they would send the appropriate signal to the interface of the LiteTouch system. A conditional statement could be written for this application. In essence, the statement would inhibit the storm shutters from lowering if any sensor was not giving an all-clear signal. In all likelihood, this would be some kind of an IF/THEN type of program. Here is a list of some conditional values (Source: LiteTouch):

(stack=x) stack is equal to value.

(stack<x) stack is less than the value.

(stack>x) stack is greater than the value.

(stack<=x) stack is less than or equal to the value.

(stack>=x) stack is greater than or equal to the value.

(stack <>x) stack is greater than or less than the value (not equal to).

Here is a conditional statement and an explanation (Source: LiteTouch):

```
GetLoadState(loadID)
If(stack=0)
{
SetLoadsOn(groupID)
}
term
```

*In this example, the program first instructs the interpreter to identify the state of the load. If the load is off, the conditional statement is true (stack=0), and the interpreter reads the "nested statement" that follows and turns on the loads in a group. A nested statement is en-*

*closed in brackets* {} *(on separate lines) and contains the action to be taken if the conditional statement is true. If the statement is false, the interpreter skips over the nested statement and interprets the statement immediately following the nested statement. In the example, if the load is already on, the conditional statement is false, so the interpreter skips the nested statement and reads the "term" command, ending the program* (Source: LiteTouch).

The next programming statement we will cover is the loop. The loop instructs the interpreter to "loop" back to the beginning of the program and start reading the program over again. If a program needs a certain function to be performed while waiting for a variable to be met, a loop is an excellent programming choice. If you use a loop in a program, you will want to make sure that there is a term statement located where it will be read sometime during the program. Without this term statement, an infinite loop will be created. As mentioned earlier, this will tie up the 5000LC processor and cause it to lock up.

A Reentrant is generally used in programs that involve timing. This option starts the program over when another input is introduced. For example, a set of water fountains turn on and off sequentially for one hour after a button press. If a reentrant is selected for this program, another button press 15 minutes after the initial will reset the program and it will now run for an hour and 15 minutes. In another timing program, a long driveway is lit from the entryway all the way to the garage. When a car enters the driveway, a sensor trips and begins the lighting program. The driveway entry lights turn on. After three seconds, the entry portion of the driveway lights turn off, then a second set of driveway lights turn on to light the next portion of the driveway. After 10 seconds, this portion of the driveway lights turn off and the next set of driveway lights turn on. If a reentrant is used in this particular program and another car drives in behind the first, the program will start over and cause the first car to travel in the dark the rest of the way. In this type of programming situation, the reentrant should be disabled. With the reentrant disabled, a second interpreter will begin the same program while the first program is still running. The reentrant option can be enabled or disabled by selecting the check-box underneath the Event field in the User Program Window. This window is where the specialty programming is accomplished.

You can access the User Program Window by selecting User Event, Startup Event, or User Function. The fields and configuration are the same for each entry selection, although the choices available in these fields may differ depending on which one you choose. When you select User Event or Startup event, there is only one option in the first field (Event). When User Function is selected, the first field will permit you to choose from three different options—Switch Press, Switch Hold, and Switch Release. Once you select one of these, a set of instructions can be written regarding the choice. The next selection is the Reentrant check-box that we discussed earlier. The program Functions list is below the reentrant check-box. Click on the program function that is required for the programming you desire. The Group, Load, Switch, or Timer fields to the right will open for you to select the load group, load switch, or timer that you want to program. Here is the complete Programs Functions List that you can choose from (Source: LiteTouch):

GetLoadState(load)—Queries whether the specified load is on (1) or off (0). The Value (0 or 1) is placed on the stack. If a scene is active that sets the load to zero, it will also return (1).

GetLoadLevel(load)—Queries for the current level (0-100) of the specified load. The value is placed on the stack.

SetLoadsOn(group)—Turns on the specified load group, either to the last-known level or to the minimum turn-on level, whichever is greater, using the load fade-on times specified in the load group parameters.

SetLoadsOff(group)—Turns off (or fades off) the specified load group.

SetLoadLevels(group)—Sets the loads in the specified load group to the preset levels defined for the group using the load fade-on times specified in the load group parameters.

SetPrevLoadStates(group)—Sets the loads of the specified load group to their previous states and levels.

GetCurrrentLevels(group)—Retrieves the level of the loads for the specified group and saves them as the new presets. This is similar to the "Scene (Custom)" function when held for six seconds.

GetMinLevels(group)—Retrieves the current level of the loads for the specified load group and saves them as the new minimum dim levels.

GetMaxLevels(group)—Retrieves the current level of the loads for the specified load group and saves them as the new maximum dim levels.

GetLoadValue(load,group)—Queries the preset value of the specified load, as a member of the specified group, and places the value on the stack.

SetLoadValue(load,group)—Uses the value at the top of the stack to set the preset level for the specified load in a specified load group.

GetLEDState(switch)—Gets the current LED state of the specified switch (0 or 1) and places it on the stack.

SetLEDOn(switch)—Turns the LED of the specified switch to "on."

SetLEDOff(switch)—Turns the LED of the specified switch to "off."

StartRamp(group)—Starts ramping the specified load group in the opposite direction of the previous ramp (knows the last ramp direction).

StopRamp(group)—Stops ramping the specified load group. This is typically programmed to switch release.

OpenLoads(group)—(Used with two-way motor loads.) Energizes the loads in a group and turns off the complements (1 on, 2 off). (Assuming the load in the group is the first output of a module.)

CloseLoads(group)—(Used with two-way motor loads.) Turns off the loads in the group and turns on the complements (1 off, 2 on). (Assuming the load in the group is the first output of a module.)

StopLoads(group)—(Used with two-way motor loads.) Turns off the loads in the group and turns off the complements.

PressSwitch(switch)—Causes a "switch press" event to be launched.

HoldSwitch(switch)—Causes a "switch hold" event to be launched.

ReleaseSwitch(switch)—Causes a "switch release" event to be launched.

ToggleLoadsOn(group)—Performs same action as the Toggle (On) function.

ToggleLoadsOff(group)—Performs same action as the Toggle (Off) function.

LockLoads(group)—Locks all loads in the specified load group to their present state. While locked they are inaccessible. They are unlocked using the "UnlockLoads" function. This program function is commonly used for parties.

UnlockLoads(group)—Unlocks all loads in the specified load group. Loads are locked with the "LockLoads" function.

LockSwitch(switch)—Locks the specified switch to its present state. While locked it is inaccessible. It is unlocked using the UnlockSwitch function.

UnlockSwitch(switch)—Unlocks the specified switch. Once unlocked, it functions normally. It is locked using the LockSwitch function.

LockTimer(timer)—Locks the specified timer to prevent it from executing. While locked, the timer can neither shut off nor turn on its load group(s). It is unlocked using the UnlockTimer function. This program function is commonly used for parties.

UnlockTimer(timer)—Unlocks the specified timer. The timer will activate at its next scheduled start time. It is locked using the LockTimer function. Note: Unlocking a timer does not cause it to start if its start time has already passed.

The next section will list the arithmetic operators and commands used in the LiteWare Programming Language. "x" represents a value (0, 1, etc.) or an identifier (load, switch, etc.). (Source: LiteTouch)

enter(x)—Enters the indicated value on the stack.

add—Adds the top two numbers on the stack together and replaces them with the result.

sub—Subtracts the last entry on the stack from the previous one and replaces both with the result.

mul—Multiplies the top two values on the stack and replaces both with the result.

div—Divides the previous value on the stack by the last entry and replaces both with the result.

mod—(Modules) When a number is divided and there is a remainder, this command places the remainder, rather than the division result, on the stack. For example:

enter(22)

enter(5)

mod

;returns the remainder of "2" to the stack.

equal—True if the top two values on the stack are equal.

nequal—True if the top two values on the stack are not equal.

gtr—True if the top value on the stack is greater than the second value.

lsr—True if the top value on the stack is less than the second value.

gtreq—True if the top value on the stack is greater than or equal to the second value.

lsreq—True if the top value on the stack is less than or equal to the second value.

setstate(x)—Sets the state variable to the value x.

wait(x)—Instructs the program to wait for the value entered in 10ths of a second (e.g., 100 = 10 seconds).

end—Causes the interpreter to stop without terminating the program.

term—Terminates the interpreter, closing the program.

pause(event,ID)—Pauses the interpreter running the specified event. The "resume" command is used to start the interpreter again. The event parameter can be represented by any of the following:
ini = startup event
swp = switch press
swh = switch hold
swr = switch release
tmb = timer begin
tme = timer end
usr = user event

The ID parameter can represent a switch, timer, or user identifier entered as a decimal value.

resume(event,ID)—Resumes a paused interpreter running the specified event. The "pause" command is used to temporarily stop the interpreter. The event parameter can be represented by any of the following:

ini = startup event

swp = switch press

swh = switch hold

swr = switch release

tmb = timer begin

tme = timer end

usr = user event

The ID parameter can represent a switch, timer, or user identifier entered as a decimal value.

getglobal(ID)—Retrieves the value from a global variable, where ID is the identifier of the global variable. A global variable is created or updated using the "setglobal(ID,val)" command. A global variable can be queried by any program on the 5000LC and will give the value that was saved to it using the "setglobal(ID,val)" command. The value is placed on the stack.

setglobal(ID,val)—Creates or updates a global variable. A global variable can be queried by any program on the 5000LC and will give the value that was saved to it using this command. The "setglobal" command saves the value (val) as the value for the specified global variable. The ID must be between 0 and 1000.

gettime—Retrieves the current time using the format HHMM (24-hour format). For example, 700 = 7:00 a.m.; 1400 = 2:00 p.m.

getdate—Retrieves the current date using the format MMDD.

getwday—Retrieves the current day of the week (1-7, where 1 is Sunday).

getmday—Retrieves the current day of the month.

Since all of the program functions, arithmetic operators, and commands have now been covered, we will now continue with more programming information.

The program function performs an action with the loads, switches, and timers. With this in mind, select the load or device that applies to the program function that you have chosen. The program will not let you choose an incompatible device/function operation, so when it denies you access to certain devices and functions, choose a different configuration.

To insert the function into the program, double-click the function. The function, group, load, switch, or timer ID will be placed into the program box. The next step will be to write the complete program into the program box using the proper commands and operators in conjunction with the program function and the load/devices. One more example of a written program will be illustrated.

*In the following example, a loop statement is used to rerun the program as long as the load level is not 50%. If the load level is set to 50% at any time, the program terminates. The program flashes the load group on and off (on one second, off one second) until the load level is set to 50% from some other source (button, timer, etc.).* (Source: LiteTouch)

```
GetLoadLevel(loadID1)
loop(stack<>50)
{
 SetLoadsOn(groupD1)
 wait(10)
 SetLoadsOff(groupD1)
 wait(10)
 GetLoadLevel(loadID1)
}
term
```

When you have completed writing a program, click Close to leave the User Program Window. This will return you to the previous dialog box. If the dialog box that you are returned to is labeled Station, click OK to save the program. To exit without saving from this dialog box, click Cancel. If the dialog box that you are returned to is either User Event or Startup, click Close to save the program. To exit without saving from this dialog box, click X.

### *Macros*

The last item that will be covered in the LiteWare section regards the creation and use of Macros. Macros are basically prepackaged programs. Usually, all you need to do is insert the specific parameters for the design into the macro that you are working with in the User Program Window. LiteTouch has macros for many different applications, and you may want to converse with company personnel before you create one yourself. It may save you some time. If you do write your own macro, be sure to save it in the LiteWare folder on your hard drive. This will make it easier to access later.

To load a macro into the User Program Window, click the Macro tab on the right side of the screen. An Open dialog box will then appear. In the LiteWare file where your macro files are kept (.mcr extension), select the one that you want to load and then click the Open tab. The Macro Parameters box will then be displayed. There are two fields in this dialog box: Parameters and Value. The Value field is a reference from the highlighted parameter field above. The Value field requires an entry for every parameter listed above. Certain parameters may need to have load groups entered into the value field. A timer parameter may need data to be entered into the value field. Once all of the parameters have had values assigned to them, click the Close tab and the macro will automatically be transferred to the User Program Window. When the complete program has been finished in the program window, click Close to exit to a dialog box. (Before you close, make sure there are LED statements included so that they will respond to the new macro.) Again, it depends on what dialog box you are in as to how you will save the macro and program information. If you are in the Station dialog box, click OK to save the program or click Cancel to exit without saving. If you are in either the User or Startup Event dialog box, click Close to save or click X to exit without saving.

The following example is a Timed (On) macro: (Source: LiteTouch).

```
@ProgramType[switch]
@DefineParam[TimedGroup,group,0]
;active time in tenths of seconds (e.g., 6000 is 10 minutes or
;600 sec.):
```

```
@DefineParam[ActiveTime(1/10sec),value,6000]
@Event[swp]
;must be reentrant (set below)
SetLoadsOn(###1)
wait(###2)
SetLoadsOff(###1)
term
@LEDUpdate[allon]
@Reentrant[1]
```

This macro only needs the load group parameter for the timer to be established, since a default time entry has been included. However, if the time needs to be different from the default, the value will have to be changed. In order for the LED update to work correctly, the load group must also be assigned to the switch (using "drag and drop"). Notice that only a switch press event is programmed and that the switch is set for reentrant. The reentrant function causes the time to start over if the switch is pressed again. (Source: LiteTouch).

The next macro example that will be illustrated is a pathway sequence (Source LiteTouch):

```
@ProgramType[switch]
@DefineParam[First Load Group,group,0]
@DefineParam[Second Load Group,group,0]
@DefineParam[Third Load Group,group,0]
@DefineParam[Delay to Turn On (1/10 sec),value,30]
@DefineParam[First Load Off Delay,value,3000]
@DefineParam[Second Load Off Delay,value,3000]
@DefineParam[Third Load Off Delay,value,3000]
@Event[swp]
;This macro sets a sequence of load groups to come on one at a time. It
;defaults to 3 seconds between each turn on and five minutes between
;each turn off.
SetLoadsOn(###1)
wait(###4)
```

```
SetLoadsOn(###2)
wait(###4)
SetLoadsOn(###3)
wait(###5)
SetLoadsOff(###1)
wait(###6)
SetLoadsOff(###2)
wait(###7)
SetLoadsOff(###3)
term
```

This macro requires load groups to be set up for each of the three steps. They are noted in the parameter setup as the "First," "Second," and "Third" load groups. There is no LED updating for the switch; therefore, no group needs to be assigned to the switch. Because the switch is not reentrant, if pressed again, the program is duplicated (Source: LiteTouch).

Here is another macro example for a dimmer (Source: LiteTouch):

```
@ProgramType[switch]
@DefineParam[Dimming Switch,switch,0]
@DefineParam[Dimming Group,group,0]
;Switch press event:
@Event[swp]
;The state variable is used to detect a hold event:
setstate(0)
end
;Switch hold event:
@Event[swh]
;The following lines cause the loads to ramp if the switch is held:
setstate(1)
GetLEDState(###1)
If(stack=0)
{
        SetLoadsOn(###2)
```

```
        wait(15)
}
StartRamp(###2)
end
;Switch release event:
@Event[swr]
;If ramping, stop the ramp, otherwise toggle the loads:
if(state=1)
{
        StopRamp(###2)
        term
}
ToggleLoadsOn(###2)
term
@LEDUpdate[allon]
```

Notice that all three events—switch press, switch hold, and switch release—have been programmed into the macro using Event tags (Source: LiteTouch).

The last macro example is a scene dimmer (Source: LiteTouch):

```
@ProgramType[switch]
@DefineParam[Scene Switch,switch,0]
@DefineParam[Scene Group,group,0]
@Event[swp]
setstate(0)
end
@Event[swh]
setstate(1)
GetLEDState(###1)
if(stack=0)
{
        SetLoadLevels(###2)
        wait(15)
```

```
}
StartRamp(###2)
end
@Event[swr]
if(state=1)
{
        StopRamp(###2)
        term
}
GetLEDState(###1)
if(stack=0)
{
        SetLoadLevels(###2)
        term
}
SetLoadsOff
term
```

In this program, if the loads are off, they fade to the preset levels when the switch is pressed and released. If the switch is held, they ramp as though assigned to a dimmer. When pressed and released again, they fade off. When turned back on, they return to the preset level using the fade-on times (Source: LiteTouch).

Some of the specialty programming may seem foreign now. In some installations, this kind of programming can solve some serious problems, so it may be worthwhile to become familiar with it. In many installations there is no need for specialty programming, so if it is not your cup of tea, it should not inhibit your progress as a fine LiteTouch technician.

This concludes the LiteWare programming and 5000LC section. There are many ways in which this type of programming environment is easier to work with than the non-Windows standard 2000 type used earlier. The advances in software will someday make even these programs seem cumbersome and hard to work with. Home automation programming is still exclusive and that makes a nice niche for the qualified people who can work with it.

# 5

# Knowing your Hardware

In this chapter, the various equipment used in the LiteTouch automated system will be reviewed. The exact function of the equipment will be discussed in addition to each of the item's specifications and operating limitations. Some of the information will already be familiar to you from previous mention during the example installation. This listing, however, will be more comprehensive.

Electronic equipment can be subject to electrostatic discharge (ESD) damage. Care should be taken when handling this equipment, especially when removing and replacing such items as EEPROM (electrically erasable programmable read only memory) chips that are contained in the LiteTouch CCU, or when handling any static-sensitive electronic components.

Surge suppression is a variable of the LiteTouch system that needs special attention. Although typical surge suppression equipment cannot defer a direct lightning strike, smaller surges can be controlled by implementing proper surge dampening hardware. Much of the LiteTouch hardware has transient spike protection built in. But, in this type of system you can really never have too much. LiteTouch can supply electrical panel units that protect the entire electrical installation. Taking these extra surge protection precautions may save some trouble down the road. Install a separate surge unit at every electrical panel that supplies power to the system. This is in addition to the main panel transient spike protector that LiteTouch will supply you with. You will probably not be sorry about the extra expenditure.

Installing copper grounding rods should improve the installation's grounding system, which in turn will improve the effectiveness of the surge suppression hardware. The telephone system is another area that could use a surge suppresser. Install a surge unit where the line enters the home. In addition, install one at the system modem or at any telephone interface location.

High- and low-voltage electrical shorts are also an enemy of the hardware system. For example, the dimmer control module outputs are not very tolerant of load shorting. Many times the output will simply not work after a load shorts to ground or between power leads. In this type of shorting situation, the rest of the remaining dimmer module outputs are usually not affected. A new load output may have to be used after shorting damage has occurred (if an extra dimmer output is available). If you do change a load output location, make sure that you change all of the switch/load programming that pertains to the new output.

The first hardware items that will be discussed are the system CCUs.

## The CCUs

LiteTouch currently has four CCUs for automation control. The Standard 2000 (which has already been discussed in length), the Compact, the Elite, and finally the 5000LC. Each CCU has unique qualities and specific applications. The two most popular CCUs are the Standard 2000 and the 5000LC. The Compact CCU can be used in smaller installations and has some limitations regarding the amount of switch stations that can be used. It also has some limitations regarding the memory available for scene presets. The Elite CCU seems to have been overshadowed by the implementation of the 5000LC CCU, so this system is currently not as prevalent. The attributes of the Standard 2000 CCU will be considered first.

### *The Standard 2000 CCU*

The Standard 2000 CCU receives all of the switch information, then processes those commands and transmits control data to the control modules. The CCU uses a nonvolatile memory for retaining programming information indefinitely. It retains clock and current load/ button press status for two days after external power is removed.

The CCU transmits and receives switch data via four station ports on the rear of the unit. These five-pin station ports are denoted by their orange color and by their designations: CSP1 through CSP4. These designations are indicated at each of their respective ports. Two of the connections on ports CSP1 through CSP4 are labeled (PS +) and (PS –), which are the power supply connections for both the CCU and the switch control stations. The voltage requirements for power supply input are between 16 and 20vdc. The amperage requirement for CCU operation is 200mA nominal. (This does not include switch station amperage requirements.) The system control modules are communicated by using the CCU's four five-pin ports labeled CMP1 through CMP4. The module communication ports and connectors are denoted by their green color and are located on the rear of the CCU.

The dimensions of the Standard 2000 CCU are 12 inches by 8¾ inches. The ambient operating temperature rating has a minimum of 45 degrees Fahrenheit (7 degrees C) and has a maximum temperature rating of 95 degrees Fahrenheit (35 degrees C).

The Standard 2000 CCU is capable of addressing 256 different control stations (2300 buttons) and up to 100 control modules (800 loads). The display screen is a monochrome LCD type. It displays a four-line by 40-character screen that measures 3 inches by 6 inches. Standard CCU 2000 grounding is accomplished from a grounding-type screw located on the rear panel.

Upgrades are achieved by opening the CCU cabinet and changing or adding IC chips. Software version numbers and identifiers are normally written on each IC for ease of change. In some CCU versions, a piggyback chip is placed over an existing IC chip for the addition of telephone voice capability.

Programming can be accomplished from the CCU's front panel that includes a 20-point touch-sensitive keypad. Programming can also be accomplished through the rear-mounted nine-pin RS232 port connector. This programming is achieved by uploading from a separate personal computer via a special serial cable. The telephone modem interface can be used to either turn system loads on or off via an off-site connection, or for off-site connecting of a PC for programming and diagnostics. This telephone interface is located on the rear of the standard CCU equipped with this optional item.

The Standard 2000 CCU employs an astronomical clock/calendar which may be used to turn system loads on or off in regard to actual daylight hours. Events may be programmed to occur on specific dates, days of the week, or on a daily basis.

The Standard 2000 CCU has no software limits on the number of masters or presets that are assigned. On/Off masters as well as scene presets may overlap load assignments with no logical conflict.

The ordering number for the Standard 2000 CCU is 08-2000-01. The ordering number for the 2000 CCU with telephone/modem interface is 08-2000-06. The order number for the Standard CCU programming software is 08-2010-01.

### The Compact CCU

Figure 5-1 depicts the wiring diagram for the Compact CCU. You can see that the control station and module wiring resembles that of the Standard 2000 CCU in many ways. Basically, the back of the Compact CCU has half of the connectors that the Standard 2000 has. The front (not shown) has three LED indicators displayed and the rest of the face is blank (no LCD display). The top LED indicator shows if the unit has power to it. The second LED down is a status indicator for proper system operation. The last LED will indicate when the modem is in use. The RS232 port is located underneath the CCU. The Compact CCU is programmed exclusively from an external PC via the serial port underneath. The same upload cable that is used for the Standard 2000 CCU can be used to upload data to the Compact. The programming software is very similar and can be accessed from the same main programs screen that was used earlier during the Standard 2000 install. The personal computer requirements for using the compact programming software are IBM compatible, 640k of RAM, 20MB of hard drive space, and DOS 5.0 or later.

This CCU has many of the same specifications in regard to current draw, ambient temperature, etc. It draws 200mA nominal and operates on 16 to 20vdc. The minimum ambient temperature is 45 degrees Fahrenheit (7 degrees C) and the maximum is 95 degrees Fahrenheit (35 degrees C).

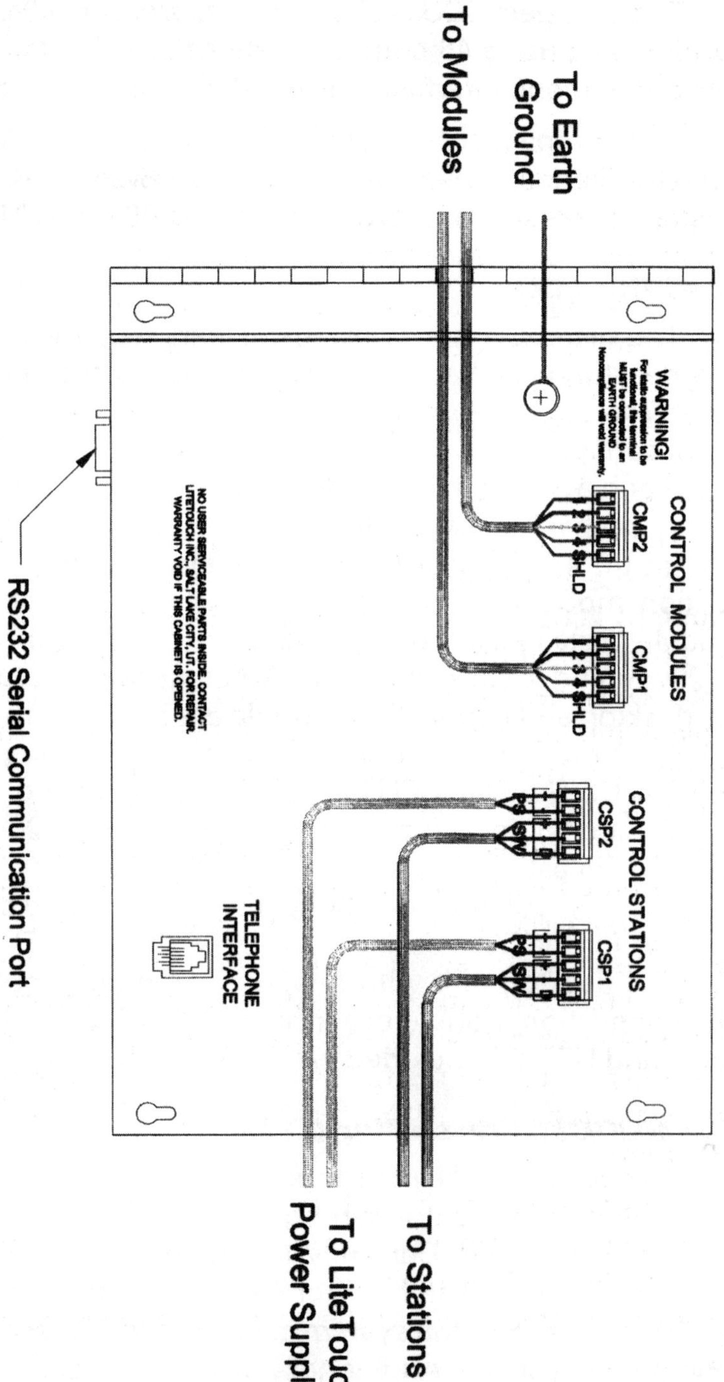
To Earth Ground
To Modules
WARNING!
CONTROL MODULES
CMP2
CMP1
CONTROL STATIONS
CSP2
CSP1
TELEPHONE INTERFACE
RS232 Serial Communication Port
To Stations
To LiteTouch Power Supplies

*Figure 5-1*

The Compact CCU is capable of addressing 40 control stations (360 buttons) and up to 40 control modules (320 loads). It can be equipped with a telephone interface and modem, just like the standard 2000.

The Compact ordering number is 08-2001-10. With a telephone and modem interface, the order number is 08-2001-16. The Compact CCU programming software order number is 08-2010-01.

### The Elite CCU

The Elite CCU is actually a two-part system. The first part consists of a desktop unit that looks very much like a standard personal computer. The desktop has a 14-inch SVGA monitor and a minitower computer configuration. In addition, it is equipped with an AT series keyboard and a 3.5-inch floppy drive (for backup). An uninterrupted power supply keeps the CCU system running during power outages. The second part of the Elite CCU system is the connection module. This connection module includes four control station ports and four control module ports located on the rear panel. Its power requirements are 16 to 20vdc and 200mA nominal. Ambient temperature ratings for both the desktop and connection module are a minimum of 45 degrees Fahrenheit ( 7 degrees C) and to a maximum of 95 degrees Fahrenheit (35 degrees C). The Elite CCU is capable of addressing 256 control stations (2300 buttons) and up to 256 control modules (2048 loads). The CCU employs an astronomical clock/calendar like that of the Standard 2000 and offers an optional telephone interface/modem capability. The order number for the Elite desktop unit is 08-2002-32. The order number for the Elite desktop unit with telephone/interface modem is 08-2002-38. The connection module order number is 08-2012-01. The interconnect cable and UPS are included with the connection module.

### The 5000LC Controller (CCU)

The 5000LC CCU incorporates a rack-mount design. Future expansion of the CCU system can be accomplished by replacing or adding new electronic cards. Currently, four cards are available: Monitor, C2000, CPU, and Modem. The Monitor card serves three purposes. It supplies power to the rest of the system cards and indicates the status of station/module bus activity. Additionally, it acts as a surge suppresser for the module bus. The C2000 card serves as the interface between the independent CPU and the 2000 equipment (control stations and modules). It can also act as a backup for the CPU card if that system fails. The CPU card stores the job programming information and does all of the processing that the system requires. It implements a nonvolatile flash memory. This card has a Y2K-compliant real-time clock that is backed up by a capacitor. The CPU was designed to be independent of the 2000 system hardware and only controls this system through the use of the C2000 card. The Modem card allows the system to be accessed by a regular telephone line. It permits voice menuing and off-site telephone control of the LiteTouch system. The modem card also allows for uploading and downloading of job programming and the use of diagnostic software called CP5000. The modem configuration is 33.6k. All four of these 5000LC system cards are extremely static sensitive. After the four cards have been inserted into the back-plane board, three future expansion slots are left unused (blanked).

The power supply input to the 5000LC is 100-240 volts ac at 50/60 Hz (dedicated feed). The output is 5vdc at 3 amps. Primary grounding must be accomplished through the grounding terminal provided (at the high-voltage input).

The CCU dimensions are 14 inches high by 16 inches wide. The depth is 5 inches. The minimum ambient operating temperature rating is 45 degrees Fahrenheit (7 degrees C) and the maximum rating is 95 degrees Fahrenheit (35 degrees C).

The 5000LC is capable of addressing 256 control stations (2304 buttons) and 256 control modules (2048 loads). Programming is accomplished through either the RS 232 serial port(s) on the CPU card or through the RJ11 jacks on the modem card. The internal back-plane board contains the four station port and the four module port connections. The 2000 system switch and load module cables are terminated to these back-plane ports. An external power supply voltage of 16 to 20vdc is also supplied to the back-plane station ports to power the system switches.

The order number for the 5000 LC enclosure with power supply is 08-5000-00. The order number for the CCU card pack is 08-5000-06. The monitor card order number is 08-5000-18. The CPU card order number is 08-5000-20. The C2000 (or LC Connection card) card order number is 08-5000-22. The Modem card order number is 08-5000-26.

## The Switches

Switch stations come in a wide variety of configurations and have many specific capabilities. Standard switches come in S, H, and E series (button and frame configurations). LTZ stations are configured to be extensions of the Data Input module. There are specialty switches like module manual overrides and slim-lines available to cater to certain client needs and construction limitations. There is also a Master Control Station that permits a visual-aid LCD screen for home automation control. Hand-held remote controls operating in the infrared range can also be implemented into the standard switch station design. All of these switches can be custom surfaced to accommodate the style and color of the interior design of the home.

### *The S-Series Switch*

This switch configuration seems to be the most popular. It permits nine buttons to be placed on its single-gang faceplate. The S-Series switch can be recognized by the small square shape of the buttons. The LED indicator is located above the button and the color green comes standard (red LED is optional). The LEDs glow with a five percent intensity when the button is turned off and glow at 100 percent when the button is turned on. The lower-intensity LED makes it easier to find the switch while the home is dark. Each switch is also equipped with a nightlight switch that can be turned on or off at the control station.

The cover plate is of a sculpted design, which means that it is a one-piece unit and that the edges are rounded off to meet the wall surface. Framed switches are available that are of a two-piece design. The earlier S-Series switch stations came standard with this framed option. You can always recognize the older-type S-Series switch stations by their LED location. The LED is illuminated beneath the button and can be seen through the button by means of a tiny window.

Two- to eight-gang switch plate configurations are available, which house two to eight separately addressed switch stations. In addition, HVAC, audio-visual, and security switches can be configured at special request.

The engraving limitations for each S-Series button are two lines at eight letters per line. The color of the engraving is usually black, but other colors like red can be specified.

When interfacing an infrared hand-held remote with an S-Series switch station, button position 9 must be used. This button position will have the IR detector mounted on the PC board below the button punch.

Each single-gang switch station is addressable. The appropriate Most Significant digit (MS or 1st) and the Least Significant digit (LS or 2nd) must be dialed in for the switch program to operate properly and for the CCU to recognize it. If a two-gang switch station configuration is used, this will have two addresses dialed in. Treat each side of the two-gang as a single switch station and this will make it easier. On a print, a two-gang switch station will have one location and two addresses assigned to it.

We will now discuss some of the specifications for the S-Series switch stations. They operate on a supply voltage of 12 to 20vdc (although 12 and 13 are rather low accepted voltages). The supply current is 10mA with no LEDs lit and 100mA with all nine lit. Specifications allow up to 35 control stations to be powered by a single power supply. From experience, try not to run this many control stations from a single power supply unless you absolutely have to. If this is necessary because of the home design, be sure to use numerous cable runs to the switches to cut down on the dc voltage drop that will be incurred. The wiring specifications call for two 18-gauge wires and one 16-gauge wire to run to the switches. A second option is three 16-gauge wires. The first specification is really a minimum and the second should almost always be opted for. Voltage drop is an inherent problem that needs to be on the mind of the cable installer. It is always best to overspecify when installing the switch station cables. If a cable run is getting too long, run another one from the CCU even if there are not many stations on that cable run. The dc voltage drops very quickly on long cable runs, especially when loaded with switch station LEDs. Low voltage tends to destroy the electronic components inside the switch stations.

The minimum ambient operating temperature for the S-Series switch station is 32 degrees Fahrenheit (0 degrees C). The maximum temperature is 100 degrees Fahrenheit (38 degrees C). The maximum number of possible addressing combinations is 256. The external dimensions for both the sculpted and framed switches are 4 5/8 inches high by 2 13/16 inches wide. The faceplates will extend from the wall about an eighth of an inch. The internal dimensions for both the sculpted and framed are 3 inches high by 2 inches wide by 1 3/8 inches deep. Square-cornered rough-in boxes that will accommodate the inside dimensions of the control stations are also specified. LiteTouch also states that plaster rings are not to be used for switch station mounting.

Each control station button can be programmed at the CCU to perform one of these many functions:

| | | | | |
|---|---|---|---|---|
| Toggle | Scene Presets | Timed Toggle | Dimming | Scene Toggle |
| Timed Flashing Interval | On Master | Raise | Momentary On | Off Master |
| Lower | Dimming Master | Toggle Master | Timed Interval | Open/Close |

The catalog ordering numbers for S-Series switch stations follow an easy pattern. For instance, we will use this order number as an example: 08-2701-XXXX. In this order number, the 27 means that it is an S-Series switch. The numbers 01 that follow the 27 mean that it is a one-button station (09 would be a nine-button). The XXXX is mostly used to denote faceplate color, bezel finish, and button color. The last X in the series denotes if the switch station will come with or without an infrared detector. Here is the current color coding list for switch ordering:

Faceplate colors:

| | | |
|---|---|---|
| (A) – Primecoat | (B) – Gray Anodized | (C) – Black Anodized |
| (D) – Polished Stainless | (E) – Satin Stainless | (K) – Satin Gold Anodized |
| (W) – White | (V) – Ivory | |

Premium Colors:

| | | |
|---|---|---|
| (H) – Polished Gold | (I) – Satin Gold | (J) – Oil Rubbed Bronze |

Bezel Color:

| | | |
|---|---|---|
| (1) – Primecoat | (2) – Black | (3) – Stainless |
| (4) – Gold | (5) – Antique Gold | |

### The H-Series Switches

H-Series switches (Figure 5-2) have rectangular-shaped buttons and, like the S-Series switches, can have up to nine buttons on a single-gang configuration. Although they are addressed in the same manner, their buttons are arranged on the faceplate quite differently. Up to a five-position switch, the buttons are aligned in a single row. The six- to nine-position configuration has two rows of buttons. The H-Series switch

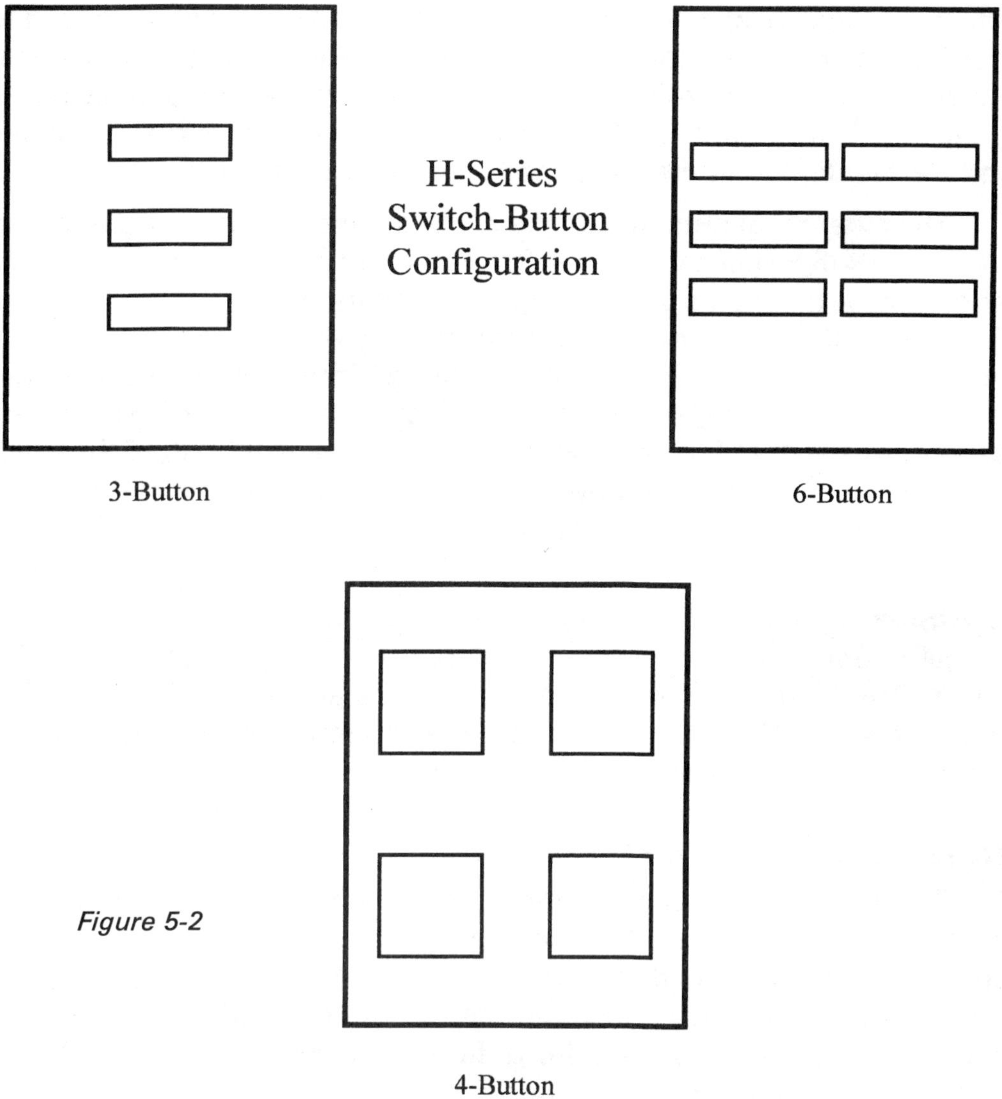


*Figure 5-2*

E-Series Switch-Button Configuration

performs in the same manner as the S-Series, so getting used to working with this type of switch does not usually present any problems. The optional infrared detector fits into the ninth button position just as it does on the S-Series. In addition, most of the other technical specifications are nearly the same on both types. One difference between them is that the engraving can only be accomplished on one line in regard to the H-Series buttons. However, the number of letters that can be used on one line increases to 10 instead of eight. This series is available in the sculpted and framed configuration, so it gives many of the same options mentioned in the S-Series specification section.

The external dimensions of the sculpted and framed H-Series faceplates are 4 5/8 inches high by 2 13/16 inches wide. The faceplates extend from the wall about an eighth of an inch. The internal dimensions for both the sculpted and framed switches are 3 inches high by 2 inches wide by 1 3/8 inches deep. The minimum operating ambient temperature is 32 degrees Fahrenheit (0 degrees C) and the maximum temperature is 100 degrees Fahrenheit (38 degrees C). All of the functions that can be programmed into an S-Series switch can be programmed into an H-Series switch.

The only difference between ordering an H-Series switch versus an S-Series switch is the need to change the first part of the number. A typical H-Series switch ordering number will look like this: 08-2901-XXXX. The 29 denotes this order number as an H-Series type (the S-series was 27). The rest of the order information can be performed in exactly the same manner as was completed in the S-Series section.

### The E-Series Switches

The E-Series switches (Figure 5-2) have much larger square buttons on their faceplates than do the S-Series. Because of this, their button configuration is different. The maximum button amount on an E-Series switch is four. These can be double-ganged to make up for the button shortage, but again, different addresses will need to be assigned to each gang that is added. Like the S-Series, the engraving schedule allows two lines at eight letters per line. The engraving letters are generally printed larger than that of either the S or H-Series. The larger engraving letter size and buttons allow people with large fingers and poor eyesight to use the system with ease. Actually, it is just a preference. Large fingers and bad eyesight probably have nothing to do with it.

As of the writing of this book, an infrared remote detector is not available for the E-Series switches. If an installation has only E-Series switches and the client wants an infrared remote installed, order an S or H-Series control station with an IR detector for the room where that capability is desired.

The rest of the specifications are nearly duplicated from either the H or S-Series switches. The minimum ambient operating temperature is 32 degrees Fahrenheit (0 degrees C) and the maximum temperature is 100 degrees Fahrenheit (38 degrees C). The order number format for the E-series is accomplished in almost the same manner as the switches that were mentioned previously. An example order number is 08-2401-XXX (some price sheets will show the fourth X). The 24 denotes the E-Series switch station. The 01 is a one-position switch. The XXX information is used for the same color input codes used for the S and H-Series switches. However, the fourth X may be missing because of the differences in remote capability.

### The Data Input Switch Station

Figure 5-3 depicts a Data Input Switch Station. This type of switch acts like two different switches combined into one. It requires two station addresses, one that is selected for the standard operating switch and one for the Data Input portion, which will respond as the next higher address. This switch connects to the standard switch station bus and outwardly appears to look just like the other switches in the series. In addition, the buttons are programmed to perform like any S, H, or E-Series. Internally, separate low-voltage cables are terminated to its data-in board to activate programmed information in the CCU. These low-voltage wires usually run to a device that gives a dry-contact closure. This can be a motion detector, photocell, door switch, or any device that closes a set of dry contacts. There is one (+V) terminal located at the top of the terminal board and one (+V) located at the bottom. From the pair of wires that run from an external dry contact device, one of the wires must go to one of these (+V) terminals. The other wire must go to one of the other terminals depending on the function of the dry contact. The first three inputs are labeled 1, 2, and 3. These are momentary-type inputs. The single-pole dry contact closure to these must close and then open again within a short period to activate the CCU program. The next set of terminals labeled 4/5, 6/7, and 8/9, are maintained contacts. When the single-pole maintained dry contacts of an external device close, a

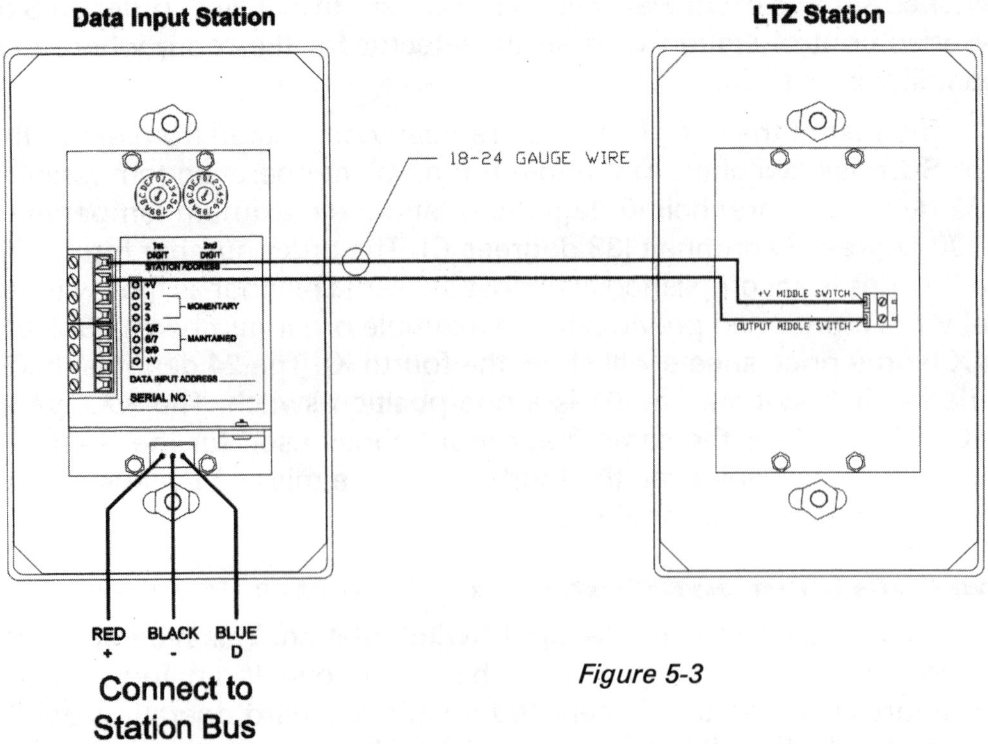


*Figure 5-3*

press and release signal is sent to the CCU. This action may be programmed to roll up a shutter, activate a door locking mechanism, or whatever is programmed into the CCU. When the maintained contact is later opened, an additional program will take effect. For example, this open-contact program may be used to unlock a door or roll down a shutter.

The wire pair or cable that runs from the Data Input Station to the dry contact location (device area) has a length limitation of 500 feet. Voltage drop must be the issue for this wire-length limitation. Since this station operates off the station bus cabling, the voltage requirements will be the same (12 to 20vdc). The lower end of this limitation is still a rather low specification and should be avoided. The current drawn by the switch will be 10mA without the LEDs lit and 100mA with nine LEDs lit. The low-voltage cabling used for the switch station bus wiring should be 16 gauge to decrease voltage drop. (16-gauge wire may be a little large for the data input terminals, so 18- to 24-gauge wires may have to be used for the separate slave inputs.) The minimum ambient operating temperature for this switch is 32 degrees Fahrenheit (0 degrees C) and the maximum temperature is 100 degrees Fahrenheit (38 degrees C). The Data Input Station is available in a sculpted or framed design and the internal and external dimensions are the same as for the S, H and E-Series switches. Contact LiteTouch for the ordering numbers for this switch.

### *The LTZ Switch Station*

In Figure 5-3 you will see the LTZ station slaved to the Data Input Station. The LTZ switch costs less than a standard addressable switch station, so it can be used in certain locations to lower the cost of the overall standard switch station expenditure. Although its outward appearance is that of a standard switch, it does not have an ID address or any "smarts". The button number limitation for each LTZ station is three. These are momentary contacts, so they need to be terminated to the Data Input Switch terminals 1, 2, and 3 (for a three-button LTZ). The cabling needed for a three-button LTZ would be a four-conductor cable (18- to 24-gauge wire). One of these four conductors is needed to travel from the (+V) terminal of the Data Input Station to the (+V) terminal in the LTZ. (Jumpers are implemented inside the LTZ to include all the +V terminals.) The remaining three conductors travel from the three outputs of the LTZ to inputs 1, 2, and 3 inside the Data Input Station. The LTZ station can also be slaved to a Data Input Module located inside the control module enclosure. Programming the data inputs controlled by the LTZ will need to be entered into the CCU after the hardwiring is accomplished.

Another LTZ-to-Data Input configuration would be three different single-button LTZ stations wired to the Data Input Station. This would consist of three separate two-wire cables entering the Data Input Station box from the three separate LTZs. The (+V) wire from each LTZ pair would terminate together and a pigtail (+V) would then run to the (+V) terminal on the Data Input Station. The other three separate LTZ cable wires would then terminate to inputs 1, 2, and 3. The termination order would be established by matching the wire colors from the proper LTZ button termination to the correct wire colors terminated at the corresponding Data Input numbers. Other LTZ wiring configurations are pos-

*Figure 5-4*

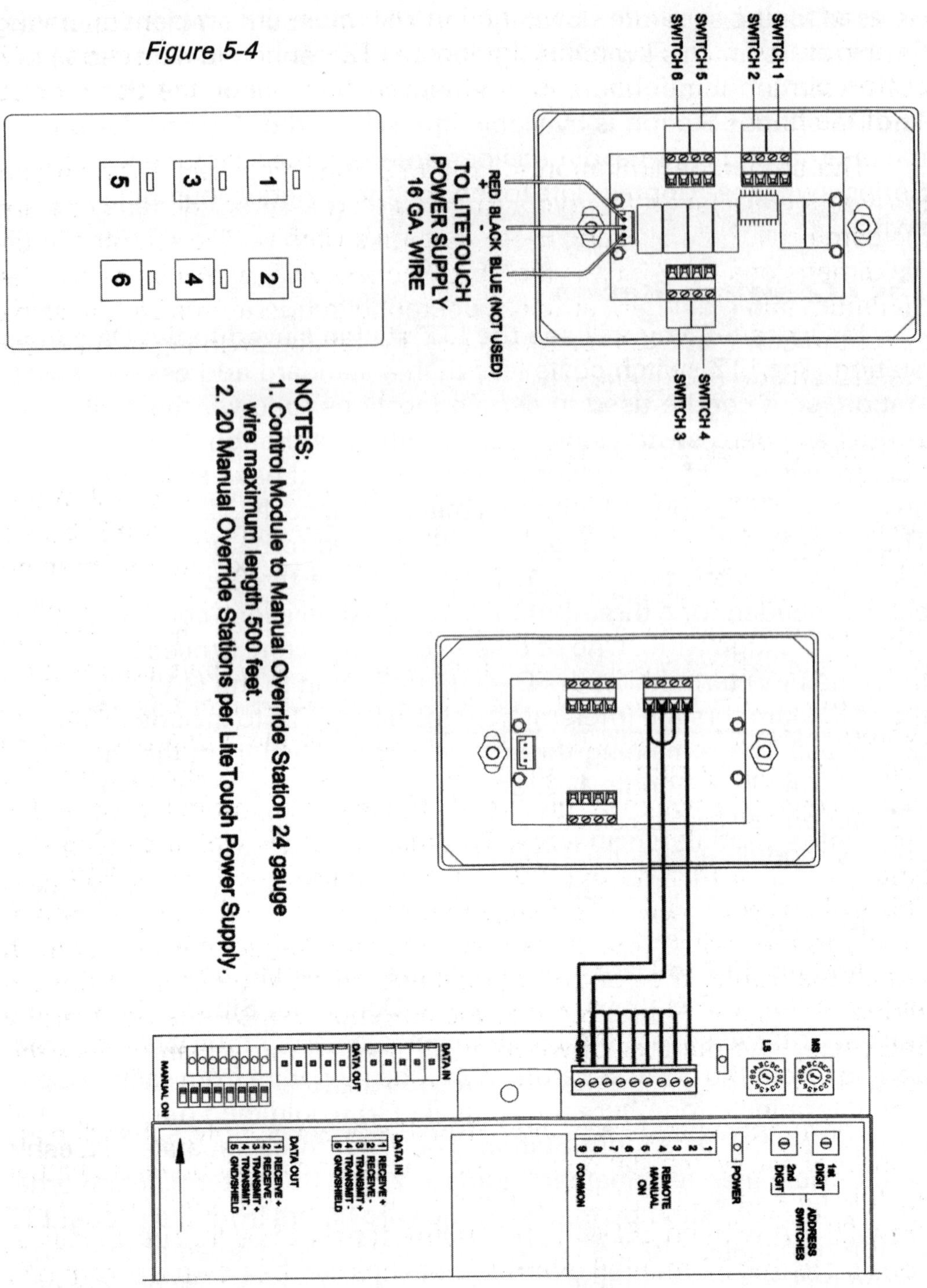


NOTES:
1. Control Module to Manual Override Station 24 gauge wire maximum length 500 feet.
2. 20 Manual Override Stations per LiteTouch Power Supply.

sible, especially if used in conjunction with more Data Input Stations or Data Input Modules. When designing an LTZ system, calculate one LTZ button per input number. The (+V) termination will be the common to all of the buttons.

The amperage limitation for the LTZ button contact is 12mA and the voltage limitation is 24vdc. The sculpted external dimensions are the same as for the S, H and E-Series-type switches. The sculpted internal dimensions are 3 inches high by 2 inches wide by ½ inch deep. The minimum and maximum ambient operating temperatures are the same as for the S, H and E-Series switches. The catalog ordering number for the LTZ station is 08-7002-01.

### Specialty Control Stations

The manual override station will be the first specialty switch discussed (Figure 5-4). This switch is similar in appearance to the S-Series switch station. Its button configuration consists of six buttons, three on each side of the faceplate. When system communications are interrupted, due to either CCU failure or control station problems, this switch can turn on the control module loads manually. This is a convenient feature for homeowners while they wait for service personnel to arrive. The system interruption does not seem as urgent to the homeowner when this switching system is implemented. The control cabling from the manual override switches run directly to the control modules located in the control module enclosures. The maximum allowable cabling distance from the manual overrides to the control modules is 500 feet. This switch requires two 16-gauge wires to run from the system power supply to the switch box. It also requires two 24-gauge wires for each switch that run to the control module enclosure. You can see from the wiring diagram in Figure 5-4 that the common can be shared, thereby decreasing the number of wires needed when more than one switch control is run to a single manual override station.

The supply voltage will come from the power supply, either directly or through the switch station bus. The amperage requirements are higher for this type of switch than for the others previously mentioned. They draw 25 mA with no LEDs lit and 100mA for six LEDs lit. The specifications state that 20 manual override switches can be powered from one

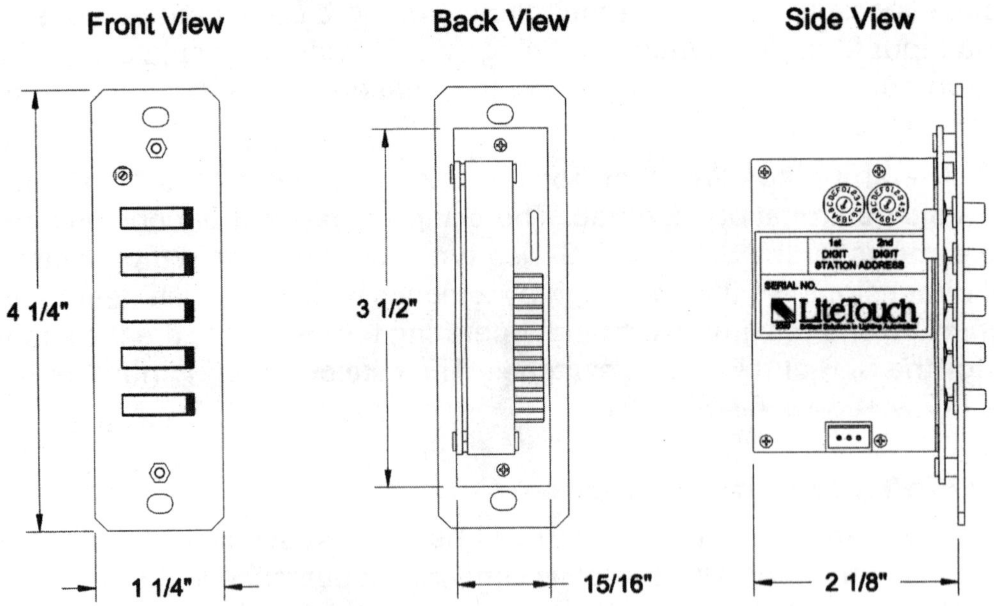


Note: Requires Slim-line Back Box LiteTouch # 08-8026-03.

*Figure 5-5*

power supply. The minimum rated ambient operating temperature for this switch is 32 degrees Fahrenheit (0 degrees C) and the maximum temperature is 100 degrees Fahrenheit (38 degrees C). The external dimensions for both the framed and sculpted switches are the same as for the S, H, and E-Series switches. The internal dimensions for both the framed and sculpted switches are 3 inches high by 2 inches wide by ¾ inches deep. The order number for this switch is 08-7006-XXX.

The next specialty switch that will be discussed is the Slim-line (Figure 5-5). This switch station measures 1½ inches wide (with cover plate). It can be useful between trim and in any area where a full-size switch station will not fit. It includes up to five H-Series buttons and comes only in the sculpted design. A special rough-in box is needed to house the switch.

The slim-line switch operates on the standard station bus voltage and draws 10mA with no LEDs lit. The internal back-box dimensions are

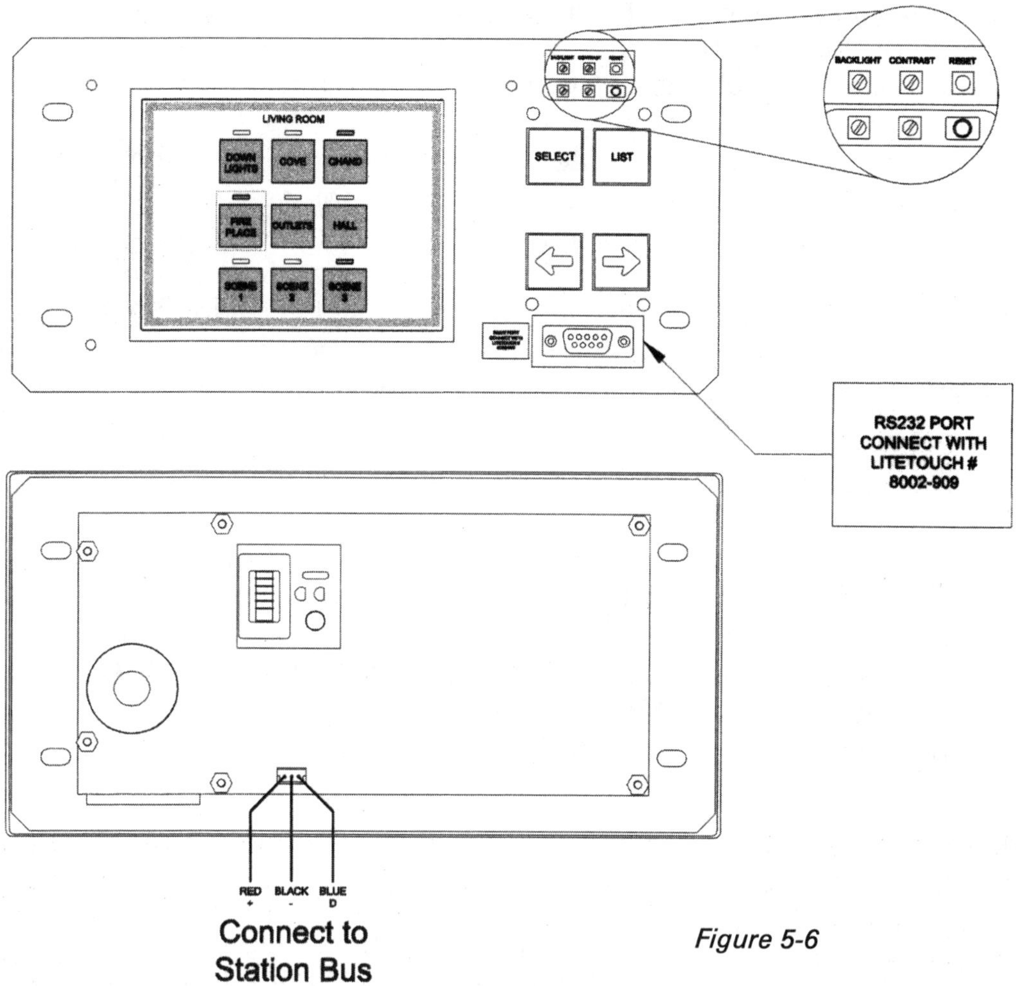


*Figure 5-6*

4 inches high by 1 inch wide by 3½ inches deep. The external dimensions are listed in Figure 5-5. The minimum ambient operating temperature is rated at 32 degrees Fahrenheit (0 degrees C) and the maximum temperature is rated at 100 degrees Fahrenheit (38 degrees C). The catalog ordering number for the rough-in box is 08-8026-03. The ordering number for the Slim-line switch is 08-7102-02.

### The Master Control Station

The master control station (Figure 5-6) will display a maximum of 32 S, H, or E-Series switches through a backlit LCD display. These displayed stations can be programmed to reveal the existing house switch stations or they can display programs that are particular only to the master station switch itself. Switch selection and control is accomplished from four E-Series buttons located to the right of the LCD screen. Programming the Master Control Station is achieved through the RS232 port located on the unit. A special software upload is completed by using the MCS Development Utility, which should be included in the standard PC software package. The overall wall space that the unit occupies is close to that of a four-gang switch plate. One power supply will drive five of these master control stations. Roughly seven control stations are equal to one master control station when calculating power consumption.

The LCD display is 340 by 240 and operates on 14 to 20vdc. Maximum current is 300mA and the power consumption is 5 watts. The minimum ambient operating temperature is 32 degrees Fahrenheit (0 degrees C) and the maximum temperature is 95 degrees Fahrenheit (35 degrees C). The back box (rough-in box) measures 4 inches high by 8.7 inches wide by 3 inches deep. The faceplate dimensions are 4.93 inches high by 9.65 inches wide by .72 inches deep. The master control station back box order number is 08-8026-08. The master control station ordering number is 08-7000-AXX.

## The Control Modules

The control modules are the interface between the low-voltage analog/digital part of the system and the high-voltage or load part of the system. Some of the modules that will be mentioned will not deal with high-voltage loads at all. They are included in this section mainly because they are installed in the same control module enclosure as that of the other modules. Special care needs to be taken when accessing the energized high-voltage-type control modules. The partition isolates the left low-voltage side from the right high-voltage side. However, it is very easy to inadvertently come in contact with the high-voltage side. High-voltage measurements and activities will be mentioned in this section to aid in the understanding of the equipment. If you are not an experienced electrician, it would be wise to contact one if you need to access the high-voltage side of the control modules. Some of the modules referred to in this section require a separate low-voltage power supply input. Some of them are self-powered from the high-voltage input located on the right side of the module.

### The Six-Channel Relay Module

The early 2000 relay modules were originally six-channel (see Figure 2-3 in the second chapter). They require a separate low-voltage input from the system power supply to operate. These six-channel control modules are installed in a great many homes, so in all likelihood, you will see them in many of the LiteTouch houses that you are working on. LiteTouch continues to sell these modules for replacement reasons. The six-channel relays do not have external fuses like that of the newer eight-channel relays. On the high-voltage side, there are two removable relay cards (three channels each). This card-removing technique can be handy when relay box contacts go bad, or when high-voltage carbon arcing has ruined a line or load terminal. In this instance, replacing only one part of the module saves the expense of replacing the entire assembly. Interlocking relay cards can also replace these standard detachable cards, effectively changing the standard six-channel relay into an interlocking type (motor control type). Six-channel relays have a few different features that separate them from the later eight-channel types, but the main difference is that they have two less high-voltage switching outputs. The minimum ambient operating temperature is 32 degrees Fahrenheit (0 degrees C) and the maximum temperature is 95 degrees Fahrenheit (35 degrees C). The overall control module dimensions are 7 7/8 inches high by 11 inches wide by 2 7/8 inches deep.

### The Eight-Channel Relay Module

This relay module is designed for dry-contact high-voltage switching of incandescent, fluorescent, switched outlet, motor, magnetic load (low-voltage) or any load that is of a nondimming type. Motor loads (120vac) should not exceed more than one-half horsepower (or 1 horsepower at 220vac). This module can control motors with higher horsepower ratings if an auxiliary relay is used in conjunction with the standard relay output. In this type of situation, the relay output would activate the coil in the auxiliary relay and the auxiliary's higher-rated contacts would then control the extra load.

The entire module (Figure 5-7) is rated at 96 amps (11,520 watts at 120vac, or 21,120 watts at 220vac). Each relay output is rated at 12 amps. For maximum contact life, however, a rating of 8.3 is recommended. Located at each of the eight outputs is an externally removable 20-amp fuse. This is a slow-blowing-type fuse and should only be changed with the module power turned off.

Power for circuit loads and for module operation is supplied by the 20-amp high-voltage power feed. The high-voltage power is connected to the top two terminals on the eight-channel relay (right side). Power jumpers are then connected from input to input until the circuit supply reaches its set load limit. This load limit is established by adding each output load of the relay and then comparing that amount to the limits of the high-voltage circuit. For instance, if a 20-amp circuit runs to an eight-channel relay and the first three channels (loads) amount to 1800 watts, then two jumpers would need to be installed starting from the first line input. The wattage amount for these three loads is under the 20-amp limit, so you may wonder why another load output is not added to this circuit. The reason for this is because the fourth channel (load) down draws 800 watts all by itself. The sum of the first three loads plus the fourth would amount to 2600 watts, so a third jumper to the fourth line input could not be included because a 20-amp circuit is not rated for this high of a wattage. The fifth channel (load) is rated at 500 watts and the sixth, seventh, and eighth are rated at 200 watts each. The sum of the wattage ratings for channels 4, 5, 6, 7, and 8 amount to 1900. This wattage amount is well within the limitations of a 20-amp circuit, so a second hot feed will be connected to the fourth input (the neutral for this circuit will go to the neutral bar). Jumpers will then be connected from the fourth input to the remaining module power inputs, thereby completing the high-voltage feeds for the entire relay module (see Figure 5-7 for a similar power jumper arrangement).

The eight-channel relay module includes a four-position interlock switch. This permits the module to either control eight loads independently or to control four pairs of interlocked loads. This interlock feature was mentioned earlier for the control of motor-type loads. A pair of relays is interlocked by turning the corresponding interlock switch to "on." This interlocking pair must then have its high-voltage line inputs

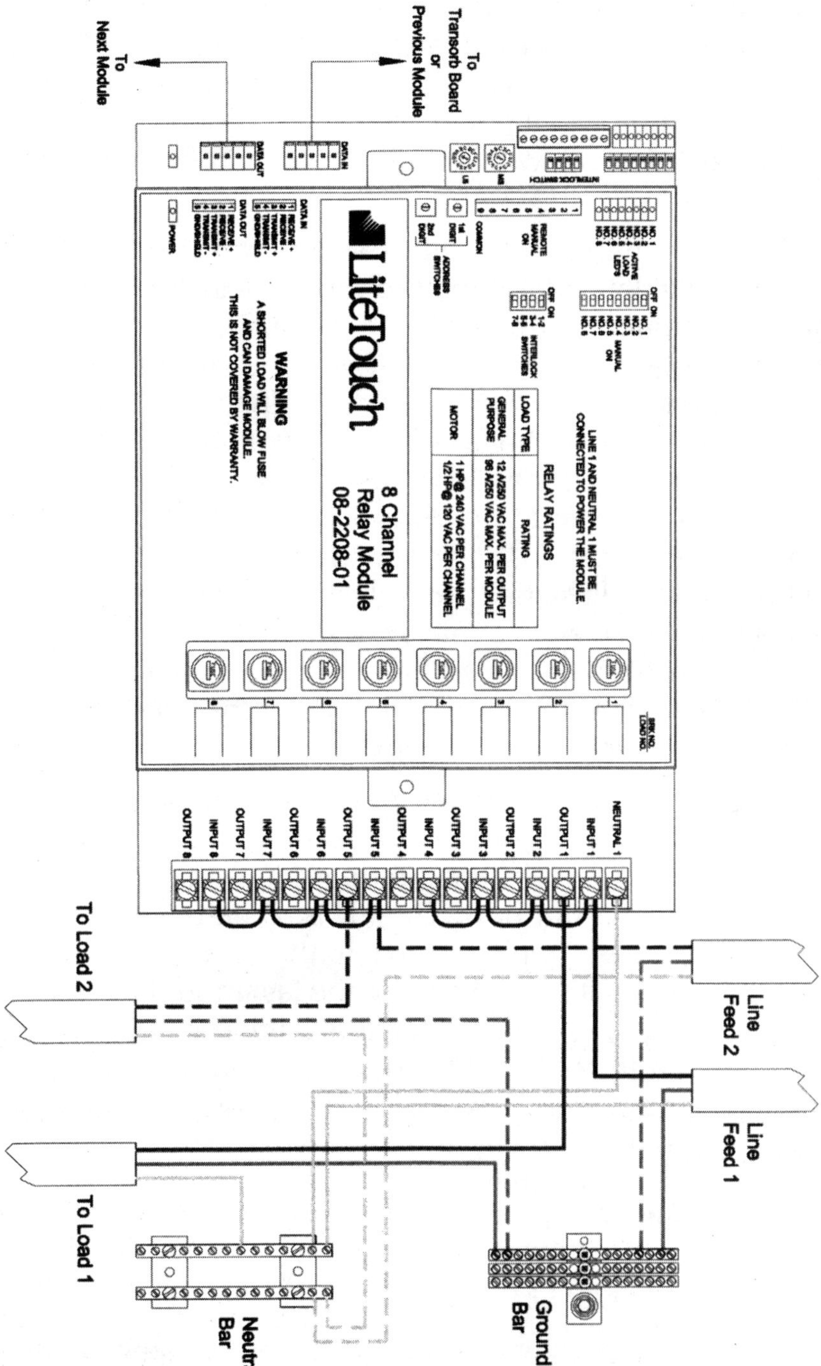
To Transorb Board or Previous Module
To Next Module
LiteTouch
8 Channel
Relay Module
08-2208-01
To Load 2
To Load 1
Line Feed 2
Line Feed 1
Ground Bar
Neutral Bar

*Figure 5-7*

jumpered together. Before you do this, make sure that the pair is supplied with only one high-voltage power feed circuit. This protection interlock helps keep power from being fed to the forward and reverse leads of the motor at the same time. Equipment that will make use of this interlock feature will include motorized skylights, shutters, entry gates, etc.

The first addressing digit or Most Significant digit is the top dial located on the left side of the module. The second or Least Significant digit is positioned just below the MS. The first number in the load address will always be the top digit and the second number in the load address will always be the lower digit. The red LED load status indicators are positioned in the far-left upper corner. These LEDs will indicate if the corresponding load is turned on and also what level the load is turned on to. In regard to relay modules, the red LEDs will either be at full brightness or off. To the right of the red LED bar, you will see a set of eight manual-on switches. You can turn each load on separately with these switches. However, make sure they are turned off for normal relay operation. The manual remote terminals are located just below the red LED bar. This is the termination point for the six-button manual remote station covered earlier in this section.

Located at the bottom left part of the module are the Data-in and Data-out ports. The top Data-in port has a communication jumper either coming in from the transorb board (from CCU) or from the preceding module's Data-out port. The lower Data-out port will have a communication jumper running from it to the next module's Data-in port, or will be left blank if it happens to be the last module in an enclosure. The first two pins (+, –) from either the Data-in or Data-out ports are for receiving data from the CCU. The next two pins (+, –) down are for transmitting data back to the CCU. The last pin is a ground/shield for the module communication system. Positioned at the far lower left part of the module is the module power LED indicator. This LED status can be a good indication that the module circuit breaker is turned on. It can also be a good indication that there may be something wrong with the module if the power LED is off and the breaker is, in fact, turned on.

Two different eight-channel relays are offered by LiteTouch—the 120vac type and the 220vac type. The first specifications will be regarding the 120vac type. Of course, the supply voltage is 120vac (60 Hz) and the module itself draws 6 watts with no load. There is an internal fuse rated at 250mA/250v. The field-serviceable fuses are rated at 20 amps/ 250v (slow-blowing). The overall module current rating is 96 amps/ 250vac maximum. The output rating is: General Purpose—12amp/ 250vac maximum, or motor—1/2 hp at 120vac; 1hp at 240vac. The contacts are specified as Form A types. The maximum module thermal dissipation rate is 7 watts (24 BTU/hour). The minimum ambient operating temperature is 32 degrees Fahrenheit (0 degrees C) and the maximum is 95 degrees Fahrenheit (35 degrees C). The overall efficiency is 99 percent at full-rated load. The overall control module dimensions are 7 7/8 inches high by 11 inches wide by 2 7/8 inches deep. The order number for this relay is 08-2208-01.

The second specification is for the 220vac eight-channel relay. The supply voltage is 220vac (50Hz) and the operating power is 6 watts at no load. The internal and field-serviceable fuses are rated the same as the 120vac unit. Maximum overall module current is 96 amps at 250vac maximum. The output rating is the same as the 120vac unit and the same Form A contacts are used. Thermal dissipation is 7 watts (24 BTU/hour). Minimum and maximum ambient operating temperature ratings remain the same as does the efficiency rating. Overall dimensions are also a duplicate of the 120vac eight-channel relay. The order number for this relay is 08-2208-02.

### *The Six-Channel Dimmer Module*

The six-channel dimmer module has also been replaced by the newer eight-channel type. These modules are installed in many homes, so there is a good chance that you will have to work with them at some time. LiteTouch continues to sell these modules for replacement reasons. This module has eight high-voltage terminals located on the right-hand side. The top terminal is for the neutral connection and the bottom terminal is for the power (hot 120vac) connection. The remaining terminals are the six load connections. This module is self-powered from the high-voltage circuit, so it does not need a separate power supply input like that of the six-channel relay. Data ports and addressing dials are configured in the same manner as other modules that have been mentioned. This unit has a Manual On switch positioned above the addressing dials that will turn on all of the module loads. It also has a Remote Manual On connection that accomplishes the same effect when it is wired to a remote switch location. Like all dimmer-type modules, the outputs are very susceptible to high-voltage shorting. This module displays a green power LED when high-voltage is supplied to the module. The rated minimum ambient operating temperature is 32 degrees Fahrenheit (0 degrees C) and the maximum temperature is 95 degrees Fahrenheit (35 degrees C). The dimensions are the same as that of the eight-channel dimmer module.

### The Eight-Channel Dimmer Module

The eight-channel dimmer module (Figure 5-8) is designed for controlling incandescent, magnetic transformer low-voltage, neon, and cold cathode loads. This unit is not recommended for use on solid-state electronic transformer low voltage. The dimmer module makes use of triacs (back-to-back silicon-controlled rectifiers—SCRs) that use phase control technology. On the new dimmers, an air gap relay is integrated into the main feed line so that when at least one load is turned on, voltage potential does not leak to the loads (triac leakage). If the dimmer module is completely off (no loads on), you can receive some rather strange readings when you are measuring dimmer output voltages. You might say that when a load is turned off there should not be any power present. However, a voltage is "usually" present at the dimmer output when all of the loads are turned off. Sometimes this measurement will float as high as a 138vac (higher than the introduced line voltage). Place a light bulb across this power and the voltage will drop. The light bulb, however, will not light. You can also get a fairly strong electrical shock from these "off" outputs if you accidentally come in contact with them. Therefore, when you measure the off voltages on dimmer modules, expect some surprising readings.

Each of the eight dimmer outputs incorporate a separate RFI (radio frequency interference) filter that is based on a toroidal inductor for low-noise operation. The overall module rating at 20 amps is 2400 watts, although it is de-rated to 1920 watts at 120vac. On the 220vac dimmer modules, the overall module rating at 20 amps is 4400 watts and is de-rated to 3520 watts. This total de-rated wattage allowance may be divided between the eight outputs in any proportion as long as no single output exceeds 15 amps or 1800 watts at 120vac. The 220vac never-

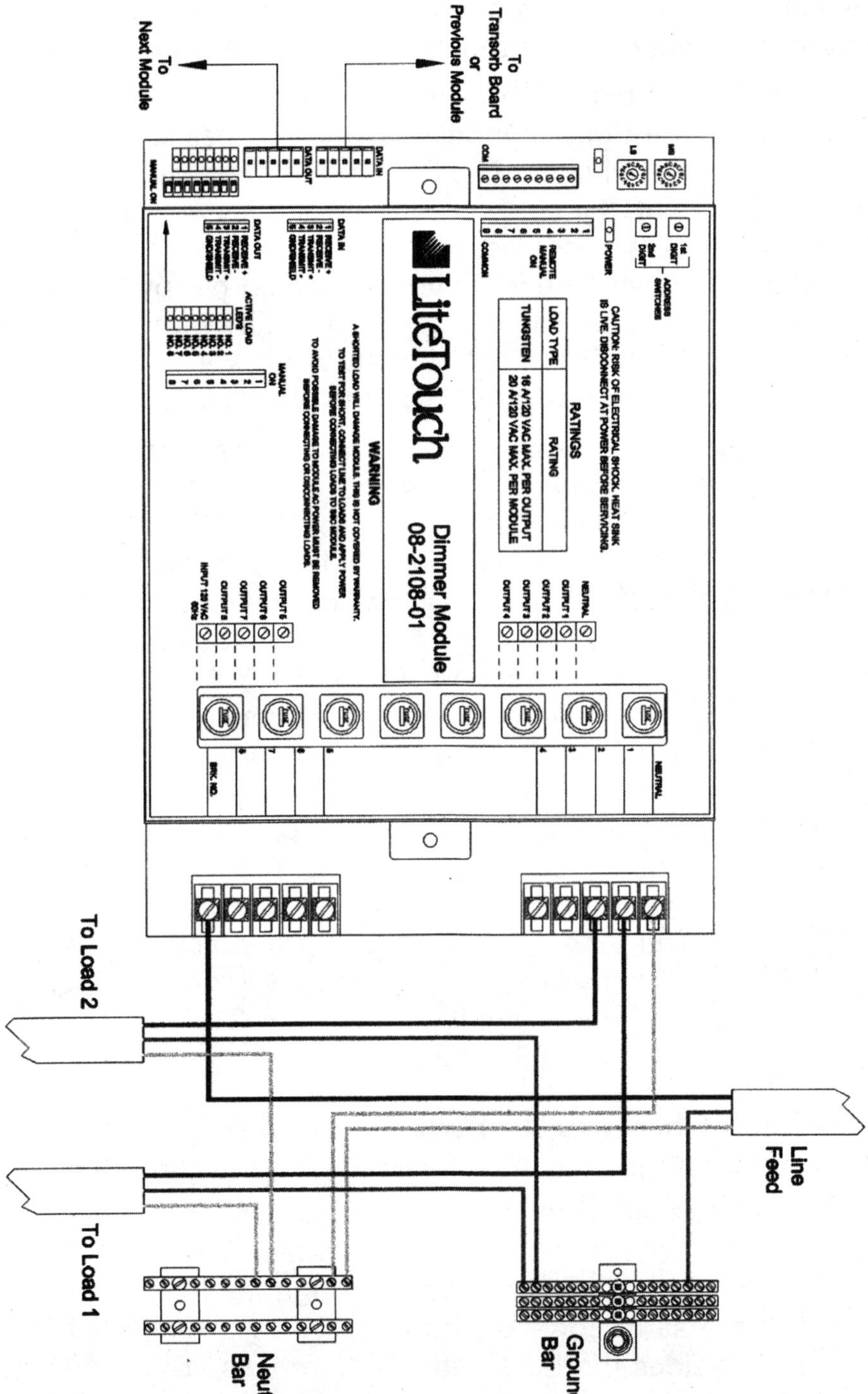
To Transorb Board or Previous Module
To Next Module
LiteTouch
Dimmer Module
08-2108-01
RATINGS
LOAD TYPE
RATING
TUNGSTEN
16 A/120 VAC MAX. PER OUTPUT
20 A/120 VAC MAX. PER MODULE
WARNING
To Load 2
To Load 1
Line Feed
Neutral Bar
Ground Bar

*Figure 5-8*

exceed single-load wattage is 3300. Each of the eight dimmer outputs has its own individual field-serviceable 15-amp fuse (slow-blow).

The low-voltage side of the dimmer module has all of the same features that the eight-channel relay module has except for the absence of the interlock switches. The Data-in and Data-out ports are located in the same area on the dimmer module as they are on the relay module. The MS and LS are located near the top-left corner and the manual-on switches are located near the bottom left corner by the LED load status indicators. The LED load indicators will dim with their corresponding high-voltage load. This is very helpful when you are at the control module enclosure and need to the see the physical ramping of the load.

The operating voltage of the eight-channel dimmer module is 120vac and the module itself uses 3 watts of power (the 220vac unit also uses 3 watts of power). The internal fuse for both modules is rated at 250mA/250v. The field-serviceable fuses for both modules are rated at 15-amp/250v (slow-blow). Overall module rating at 20 amps is 2400 watts de-rated to 1920 watts (120vac). The 220vac module rating at 20 amps is 4400 watts de-rated to 3520 watts. The individual output rating for the 120vac module is 15 amps at a maximum of 1800 watts. The output rating for the 220vac module is 15 amps at a maximum of 3300 watts. Maximum thermal dissipation rate for both modules is 25 watts (84 BTU/hour). The minimum ambient operating temperature rating for the 120vac and 220vac units is 32 degrees Fahrenheit (0 degrees C). The maximum temperature rating for both is 95 degrees Fahrenheit (35 degrees C). The external dimensions are exactly the same for either module—7 5/8 inches high by 10¼ inches wide by 3 inches deep. The order number for the 120-volt module is 08-2108-01. The order number for the 220-volt unit is 08-2108-02.

### The Quad Dimmer Module

This quad dimmer module (Figure 5-9) provides two 20-amp dimmers in a standard size module enclosure (four output loads). Like the eight-channel dimmer unit, it incorporates a separate RFI filter for low-noise operation. It is addressable through the MS and LS dials and has most of the same low-voltage featured in the eight-channel dimmer module. Besides the load output differences, it varies from the eight-

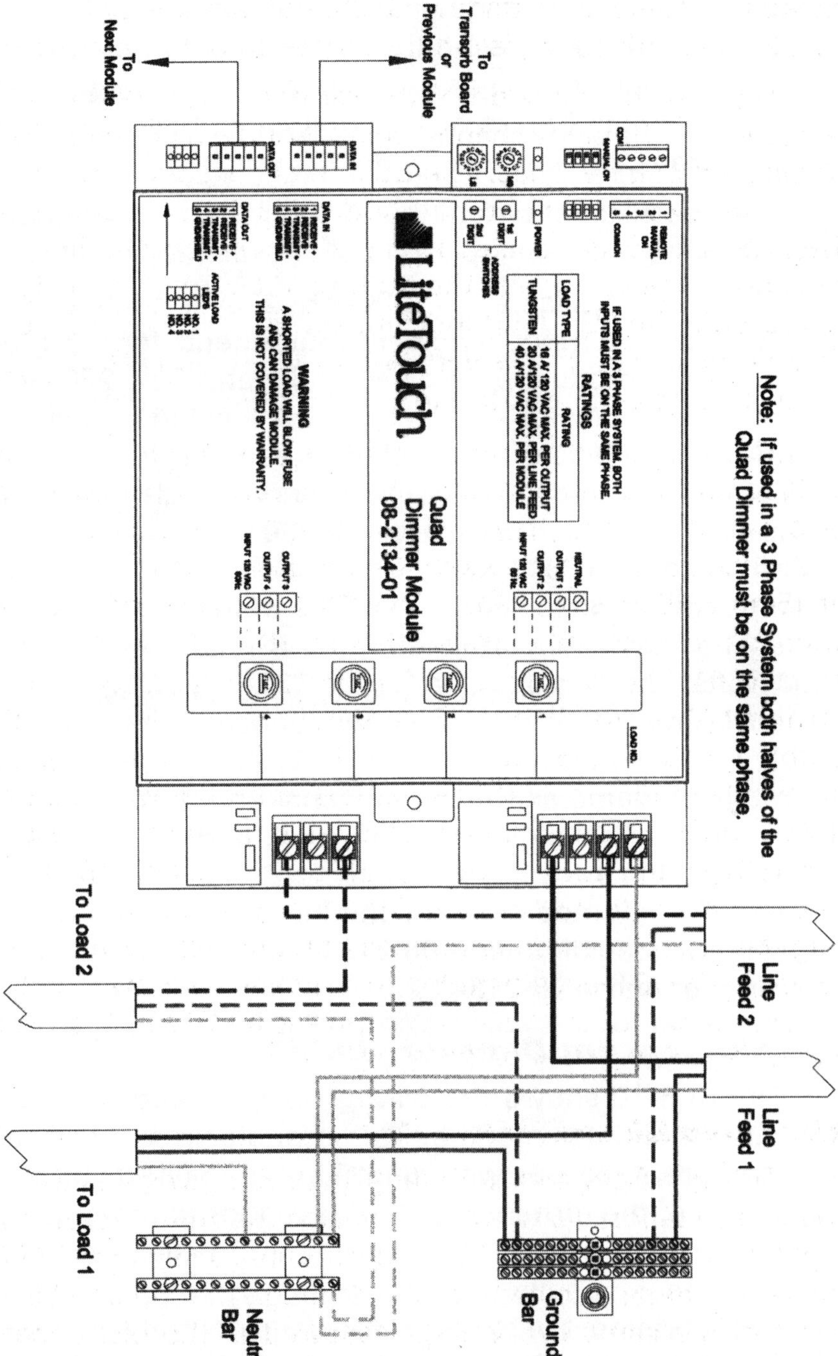
To
Transorb Board
or
Previous Module
To
Next Module
Note: If used in a 3 Phase System both halves of the
Quad Dimmer must be on the same phase.
LiteTouch
Quad
Dimmer Module
08-2134-01
WARNING
A SHORTED LOAD WILL BLOW FUSE
AND CAN DAMAGE MODULE.
THIS IS NOT COVERED BY WARRANTY.
Line
Feed 2
Line
Feed 1
To Load 2
To Load 1
Ground
Bar
Neutral
Bar

*Figure 5-9*

channel dimmer in that it has two high-voltage power inputs. These power inputs need to be on the same phase when using three-phase power. Special instructions are required when using GFI (ground fault interrupter) circuit breakers. This unit cannot be used to control solid-state electronic low-voltage loads.

The specifications state that if 550 ms rise times are required, 10-amp, 15-amp, or 20-amp peripheral chokes may be purchased.

This unit is available in a 120vac or 220vac configuration. The operating voltage and module power requirements for the 120-volt unit are 120vac with a module power use of 3 watts. The 220vac module also uses 3 watts of power. The internal fuse located in each unit is rated at 250mA/250v. The field-serviceable fuses (four) for both units are rated at 15 amps/250v (slow-blow). The overall module rating for the 120-volt unit is 40 amps, which calculates to 4800 watts. This calculated wattage is then de-rated to 3840 watts. The module rating for the 220-volt unit is 40 amps calculated to 8800 watts. This wattage is then de-rated to 7040 watts. The individual output rating for the 120-volt unit is 15 amps and 1800 watts. The output rating for the 220-volt unit is 15 amps and 3300 watts. Maximum thermal dissipation for both units is 40 watts (135 BTU/hour). The minimum and maximum ambient operating temperatures are rated the same as the eight-channel relay and eight-channel dimmer modules. The efficiency rating is 99 percent at full-rated load. The dimensions for the quad dimmer module are the same as the eight-channel dimmer module mentioned previously. The order number for the 120-volt quad dimmer is 08-2134-01. The order number for the 220-volt dimmer unit is 08-2134-02.

### The Fluorescent Dimmer Module

Figure 5-10 shows the wiring diagram and the drawing for the fluorescent dimmer module. This module was specifically designed for fluorescent fixtures with dimming-type ballasts. It has four output pairs. One of the terminals of the pair is for dimming and the other is for the constant output that the dimming ballast requires. The high-voltage feed is supplied at the top and bottom terminals. In the 120-volt configuration, the top terminal is the neutral and the bottom is the hot terminal. In the 220-volt configuration, both terminals are hot. This module is equipped with RFI filters for noise reduction and field-ser-

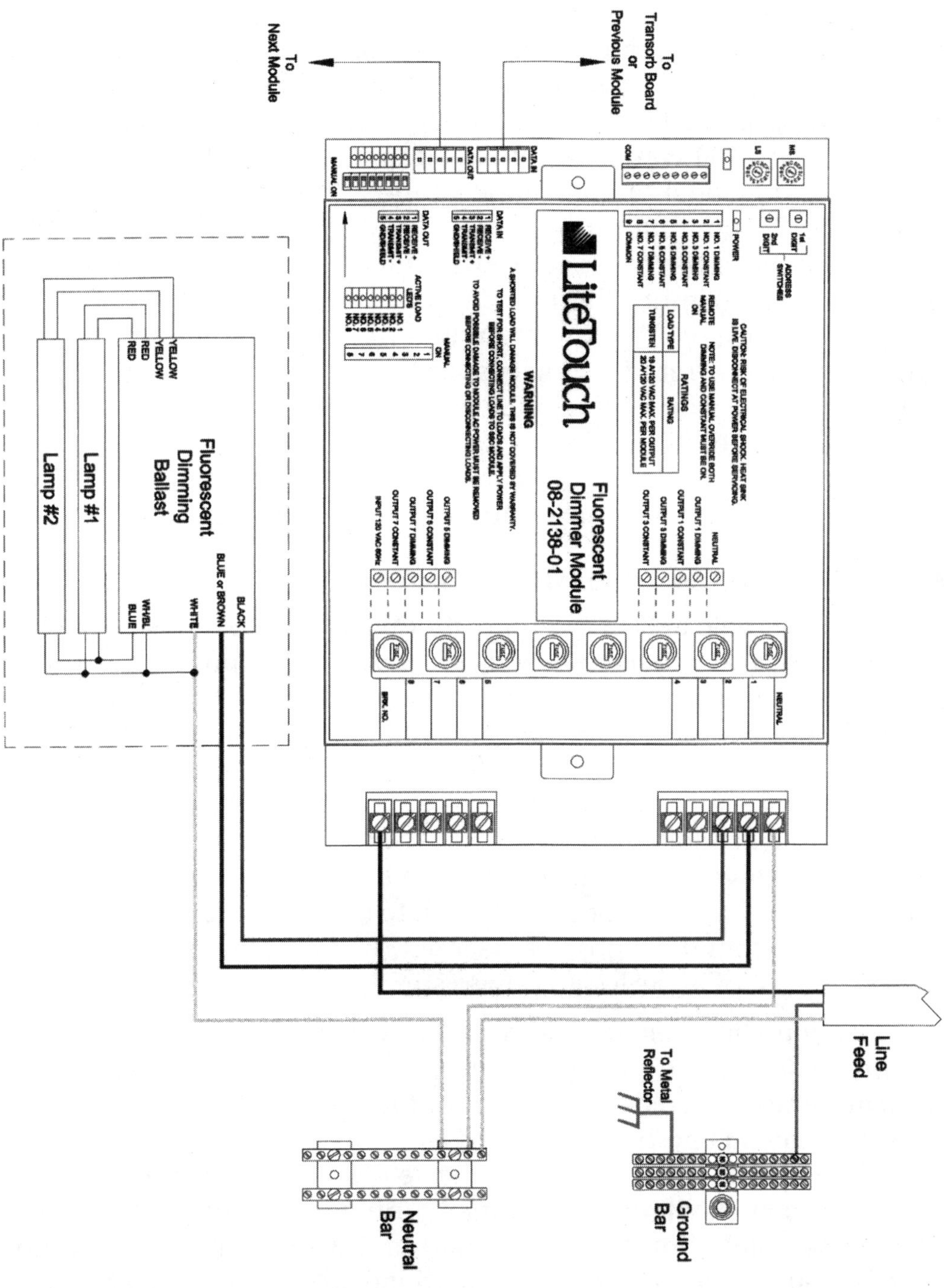
To Transorb Board or Previous Module
To Next Module
LiteTouch
Fluorescent Dimmer Module 08-2138-01
WARNING
Fluorescent Dimming Ballast
Lamp #1
Lamp #2
YELLOW
YELLOW
RED
RED
BLACK
BLUE or BROWN
WHITE
W/BL
BLUE
Line Feed
To Metal Reflector
Ground Bar
Neutral Bar

*Figure 5-10*

viceable fuses for both the dimming and constant outputs. In each voltage configuration the module power usage is 3 watts. Both have an internal fuse rated at 250mA/250v. The field-serviceable fuses are rated at 15 amps/250v (slow-blow). The 120-volt unit's module rating is 20 amps at 2400 watts. The 220-volt unit rating is 20 amps at 4400 watts. The individual output rating of the 120-volt unit is 15 amps at 1800 watts. The output rating of the 220-volt unit is 15 amps at 3300 watts. They both have four outputs with an accompanying filament channel. The maximum thermal dissipation rate is 25 watts (84 BTU/hour) for either unit. The minimum and maximum ambient operating temperatures are rated the same as the other modules previously mentioned. The overall efficiency is 98 percent at full-rated load and the module dimensions are the same as the dimmer modules mentioned previously. The 120-volt module order number is 08-2138-01 and the 220-volt module order number is 08-2138-02.

### Electronic Ballast Dimming Module

The electronic ballast dimming module (Figure 5-11) is a six-channel controller that provides a zero- to 10-volt dc output that is directly proportional to a selected lighting level. In addition, this module provides six dry-contact relays for high-voltage switching. The contacts will close whenever the corresponding analog output is above zero volts dc. This module was made to operate fluorescent ballasts that need an analog output in addition to switched 120vac. Each analog output of the module can drive about 10 ballasts. This is one of the modules that needs a low-voltage power supply input to operate. The module itself is not powered from the high-voltage input. This module uses about one-fifth the output of a LiteTouch power supply.

The operating voltage and power requirements are 14 to 20vdc at 350mA (no load). The supply voltage is 16 to 20vdc. The internal fuse is rated at 1 amp/250 volts. The analog output rating D/A resolution is 250 steps and the analog output drive is zero to 10 volts proportional. The analog output drive rating per channel is 4½ mA (about eight ballasts). The specifications state that all analog outputs are protected against short circuit and short-term miswiring to 120vac. If additional output drive is required (up to 20mA), the miswiring protection can be removed.

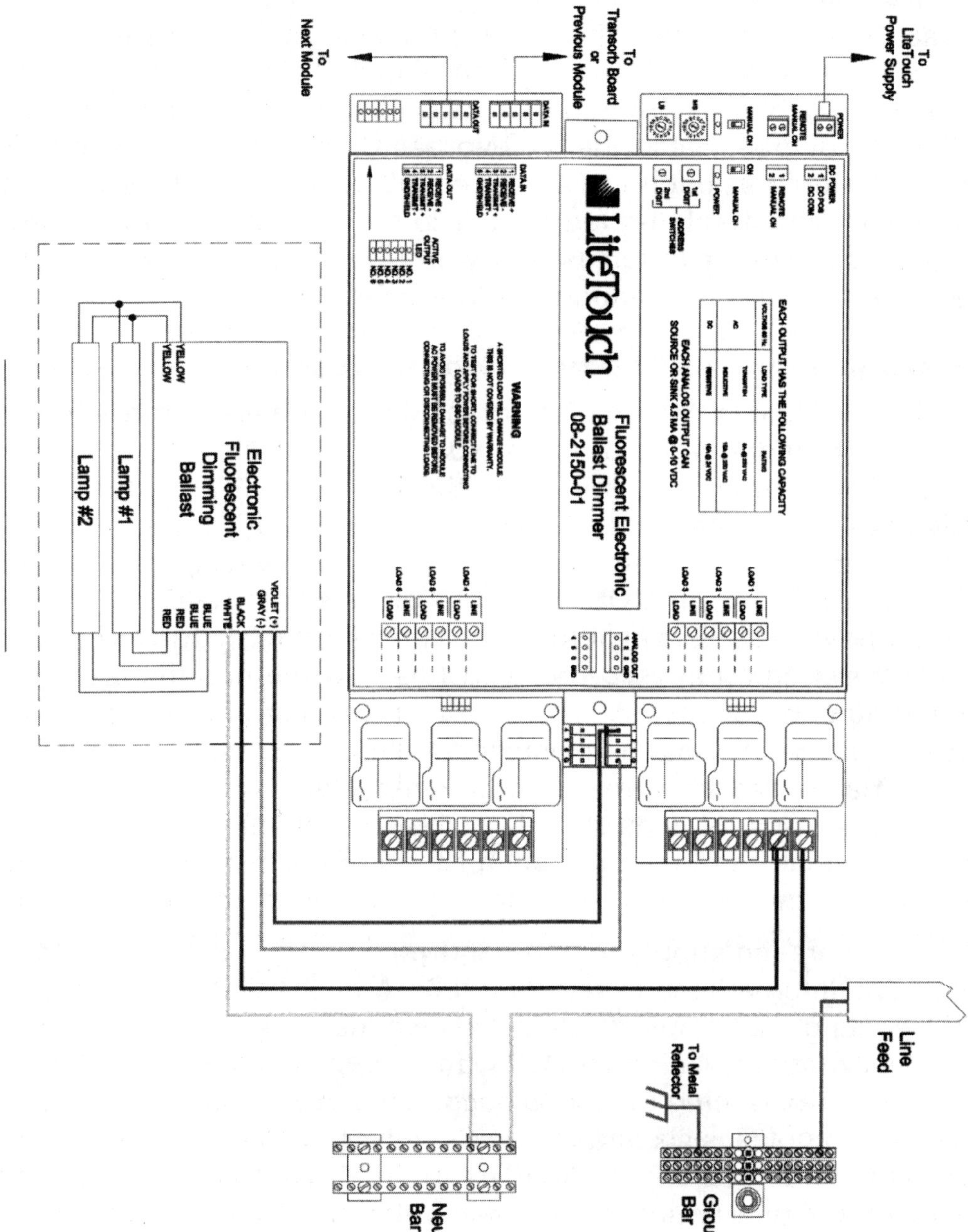
To LiteTouch Power Supply
To Transorb Board or Previous Module
To Next Module
LiteTouch
Fluorescent Electronic Ballast Dimmer
08-2150-01
WARNING
Electronic Fluorescent Dimming Ballast
Lamp #1
Lamp #2
YELLOW
YELLOW
VIOLET (+)
GRAY (-)
BLACK
WHITE
BLUE
BLUE
RED
RED
Line Feed
To Metal Reflector
Neutral Bar
Ground Bar

*Figure 5-11*

The relay output contact rating for tungsten is eight amps at 250vac. For inductive loads, it is 15 amps at 250vac. For resistive loads, it is 15 amps at 24vdc. The maximum thermal dissipation rate is 10 watts (34 BTU/hour). The minimum ambient operating temperature rating is 32 degrees Fahrenheit (0 degrees C) and the maximum temperature rating is 95 degrees Fahrenheit (35 degrees C). The efficiency is over 98 percent at full-rated load. The module dimensions are the same as the dimmer modules that were mentioned previously. The order number for this dimming module is 08-2150-01.

### Electronic (Inverse Phase) Dimmer Module

Figure 5-12 depicts the electronic (inverse phase) dimmer module. This module was designed for solid-state transformer low-voltage lighting. Sometimes the triac (the component LiteTouch normally uses for dimming) can cause excessive lamp filament noise or shorten the life of the electronic transformer. However, this module incorporates the power MOSFET (metal oxide silicon field-effect transistor). These MOSFETs produce smoother switching due to the fact that they switch on at zero-crossing (they switch on whenever the alternating current wave crosses the zero voltage point). This zero-crossing allows the load current to rise slowly at the normal 60 Hz range. Triac dimmers switch at mid-half-cycles, and this generates steep waveforms. This module is not recommended for inductive-type loads such as magnetic transformers or motors (loads with internal windings).

The operating/supply voltage and power rating are 120vac at 7 watts. The internal fuse is rated at 250mA at 250 volts. The overall module rating is 20 amps at a de-rated wattage of 1920. The individual output maximum is 4 amps at 480 watts. Since this single-output rating is so low, balancing the loads becomes more important. The total number of outputs is six and the unit's maximum thermal dissipation rate is 41 watts (138 BTU/hour). The rated minimum and maximum ambient operating temperatures match the values of the previously mentioned modules. The efficiency rating is over 98 percent at full-rated load. The dimensions are the same as the other dimmer modules mentioned earlier. The electronic (inverse phase) dimmer module ordering number is 08-2140-01.

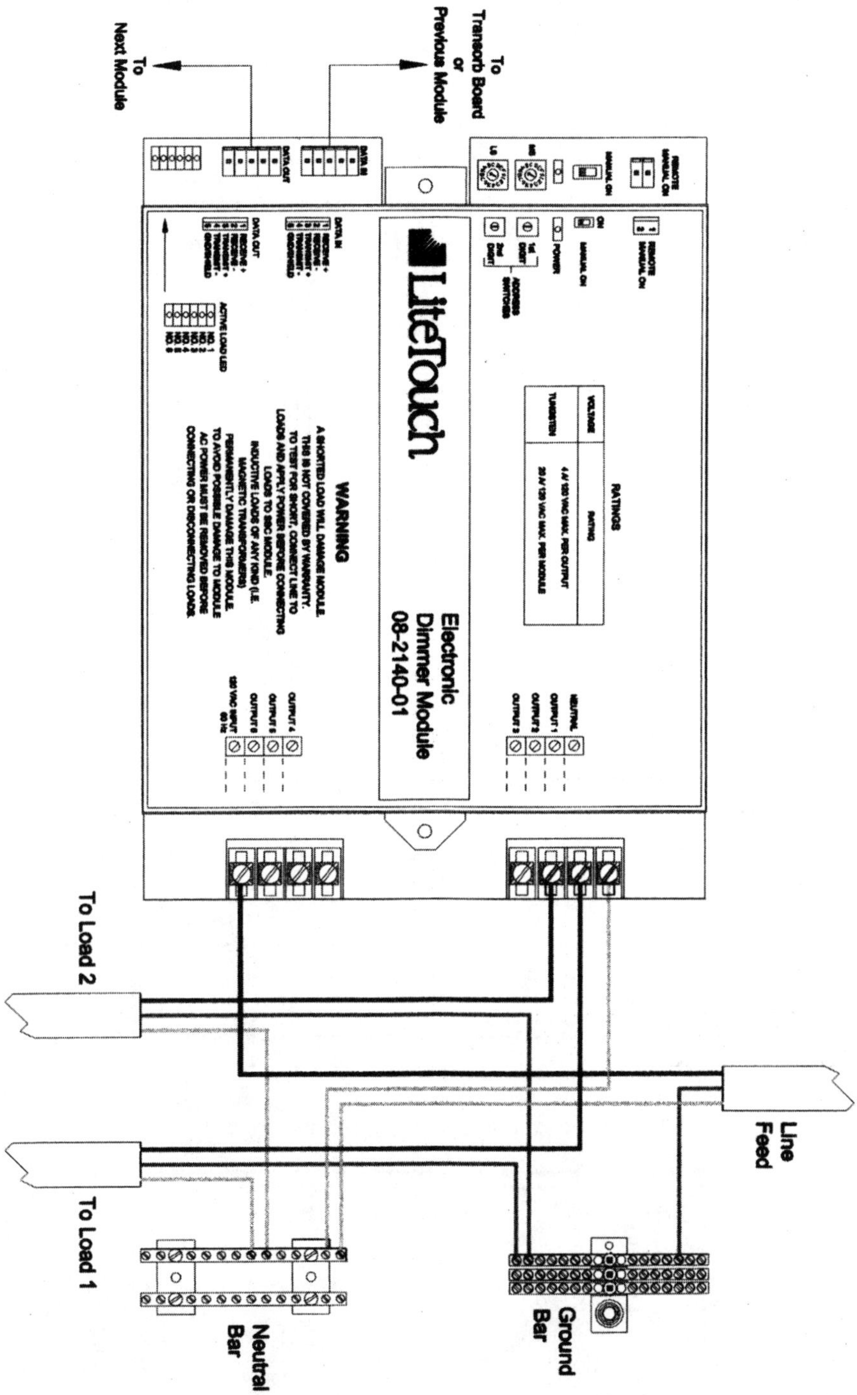
To Transorb Board or Previous Module
To Next Module
LiteTouch
Electronic Dimmer Module 08-2140-01
WARNING
RATINGS
To Load 2
To Load 1
Line Feed
Ground Bar
Neutral Bar

*Figure 5-12*

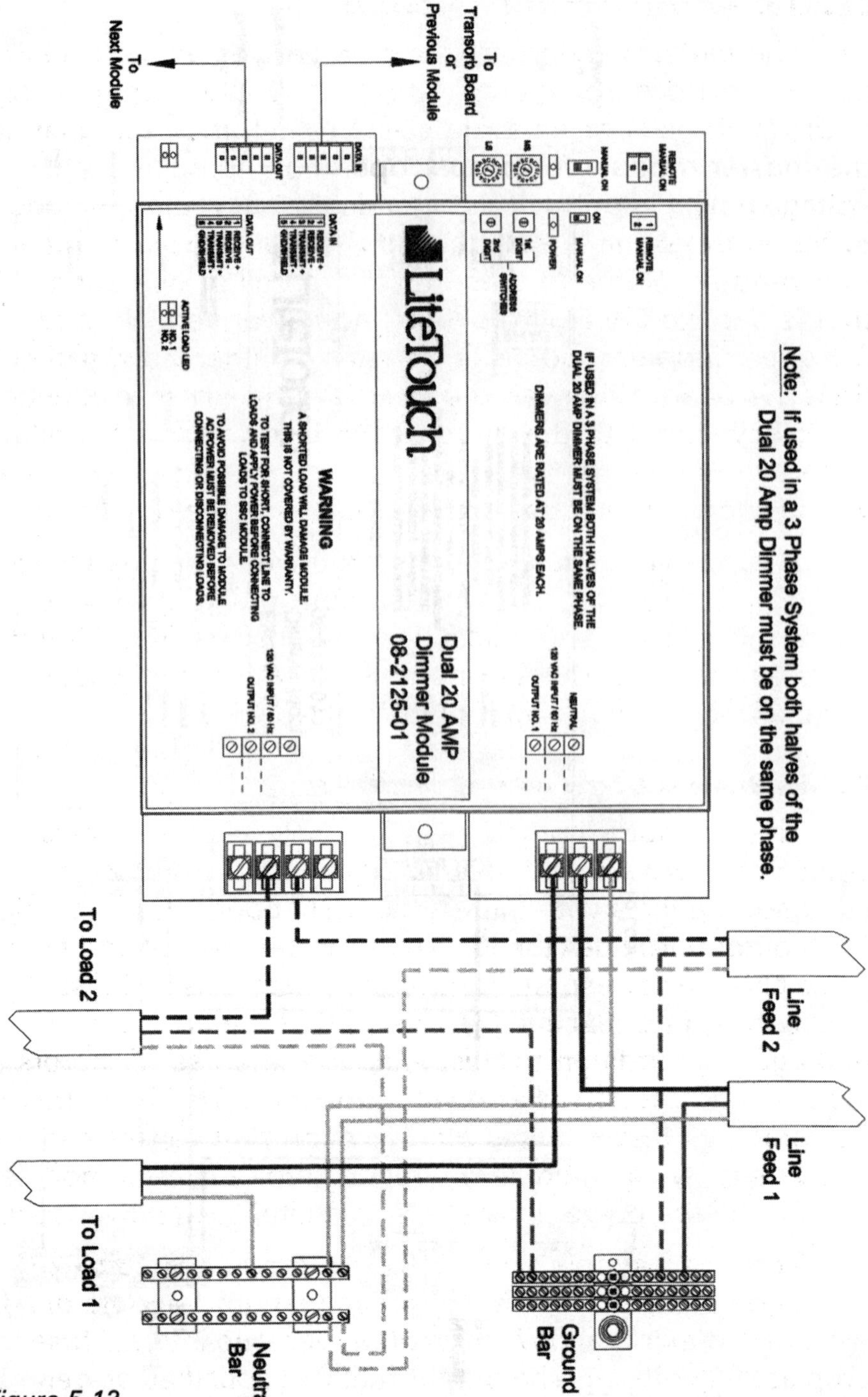

To Transorb Board or Previous Module
To Next Module
Note: If used in a 3 Phase System both halves of the Dual 20 Amp Dimmer must be on the same phase.
LiteTouch
Dual 20 AMP Dimmer Module 08-2125-01
WARNING
Line Feed 2
Line Feed 1
To Load 2
To Load 1
Ground Bar
Neutral Bar


*Figure 5-13*

### Dual 20-Amp Dimmer Module

The earlier version of this module was actually a slave dimmer controlled from a standard dimmer module output. It did not have addressing dials or data ports and relied upon the analog output of the master module for proper operation. In addition, the input high-voltage power supply to this older dual 20-amp dimmer module needed to be on the same leg (phase) as the input power of the master dimmer module. Some of these units are still in operation. The module that is depicted in Figure 5-13 is the newer version. It competes with the quad dimmer module since it has a higher individual output rating than the quad. However, the dual 20-amp dimmer only controls two loads compared to the quad's four. Like the quad, it allows for two separate load feeds. When using three-phase power, make sure that you use the same phase on both halves of the dimmer module.

The overall module rating is 4800 watts and the individual output rating is 1920 watts. This single-output rating effectively de-rates the entire module to 3840 watts. The dimensions of this module are the same as the other dimmer modules mentioned previously. The order number for this dimming module is 08-2125-01.

### DC Motor Control Relay Module

The dc motor control relay module (Figure 5-14) is designed to control the forward and reverse leads on a dc motor-type load. It contains six channels with three separately operated dry-contact outputs. Each output provides the terminals for the positive and negative motor leads plus the corresponding power connections for these. The dc motor loads should not exceed 5 amps so that the control relays will not be damaged. In addition to this limitation, the use of motors with noise suppression circuitry is also recommended. This is another of the control modules that requires a connection to a separate low-voltage dc power supply for module operation. Typically, one module will draw one-fifth of the power limit of a single LiteTouch power supply.

The rated supply voltage to this module is 16 to 20vdc and the operating current draw is 350mA maximum (module only). This unit will use a maximum of 7 watts of power. The internal fuse is rated at 1 amp at 250 volts ac. The output rating is specified as general purpose

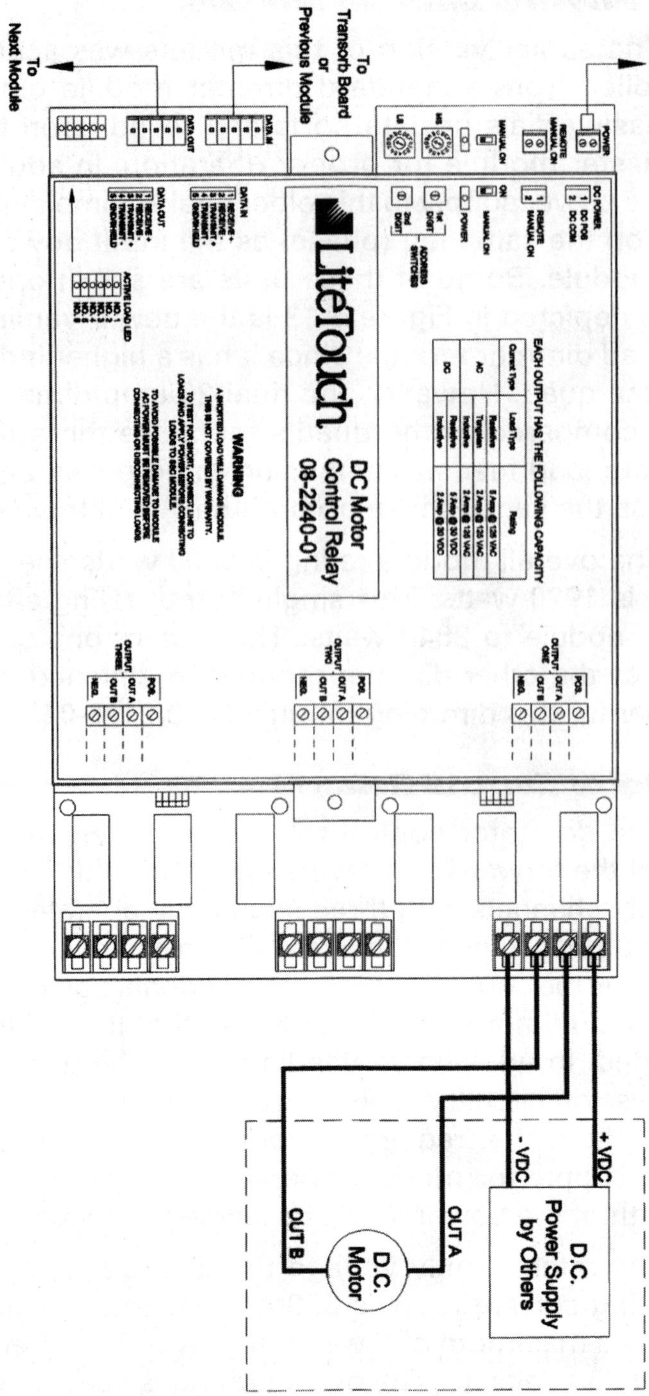
To LiteTouch Power Supply
To Transorb Board or Previous Module
To Next Module
LiteTouch
DC Motor Control Relay 08-2240-01
WARNING
EACH OUTPUT HAS THE FOLLOWING CAPACITY
+VDC
- VDC
D.C. Power Supply by Others
OUT A
OUT B
D.C. Motor

*Figure 5-14*

5 amps at 30 volts ac or dc. The motor rating is one-sixth horsepower at 24vdc. The contacts are a combination of Form A and Dual Form C (dedicated to dc motor control). The rated minimum operating ambient temperature is 32 degrees Fahrenheit (0 degrees C) and the maximum rated temperature is 95 degrees Fahrenheit (35 degrees C). Module efficiency is rated at over 99 percent at full-rated load. The overall dimensions are the same as the modules that were mentioned previously.

### Low-Voltage Eight-Channel Relay Module

The low-voltage eight-channel relay module (Figure 5-15) closes sets of low-voltage dry contacts for controlling such items as HVAC, audio/video, alarms, and some window treatment controls. The uses in home automation seem endless with this type of module. Virtually any electronic or home appliance can be controlled through this module with the proper interfacing techniques.

The low-voltage contact ratings are limited to 3 amps at 24 volts ac or dc and are located on the left side of the module. This is another module that requires a low-voltage power supply input to operate. Again, it will use about one-fifth of the power of a LiteTouch single power supply. The module operating voltages and power requirement are 14 to 20vdc at 330mA. The internal fuse is rated at 1 amp at 250 volts. The low-voltage contacts are Form A, and the maximum module thermal dissipation rate is 6 watts (20 BTU/hour). The rated minimum operating ambient temperature is 32 degrees Fahrenheit (0 degrees C) and the maximum temperature rating is 95 degrees Fahrenheit (35 degrees C). The dimensions are the same as the previous modules mentioned. The low-voltage relay module ordering number is 08-2235-01.

### Analog Output Module

This is a six-channel control device that provides a zero to 10-volt dc output that is directly proportional to the desired lighting level. This module is used as an interface to remotely located special dimmers. These special dimmers are required for dimming high-power incandescent and other assorted gas-discharge lamps. Each line of analog output can drive five standard remote dimmers.

*Figure 5-15*

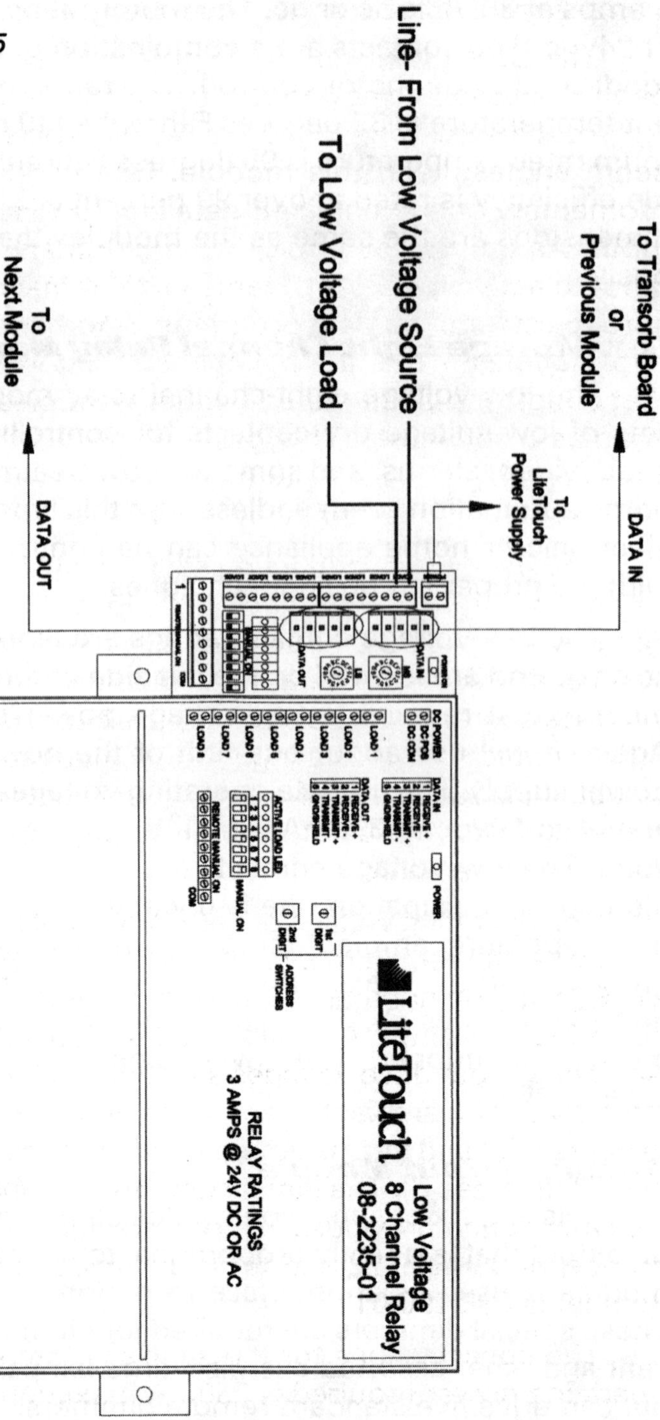
To Transorb Board
or
Previous Module
DATA IN
To
LiteTouch
Power Supply
Line- From Low Voltage Source
To Low Voltage Load
To
Next Module
DATA OUT
LiteTouch
Low Voltage
8 Channel Relay
08-2235-01
RELAY RATINGS
3 AMPS @ 24V DC OR AC

### ***Data Input Module (Momentary and Maintained)***

The Data Input module (Figure 5-16 shows the momentary) has been discussed throughout the book as an interesting and useful interface. Like the low-voltage relay, the home automation possibilities seem endless with this module. One of the contrasts between the momentary and maintained data input modules is the function of the input. As the names imply, the momentary requires a short contact input to activate CCU programs and the maintained requires a set closure to activate the CCU programs. Another difference between these two modules is that the maintained data input has four DIP switches that couple inputs together. Switch 1 couples inputs 1 and 2. Switch 2 couples 3 and 4, etc. This coupling allows the pair to work together. For example, the first input may be an "on" function for a set of loads and the second input may be an "off" for the same set of loads.

The data input module is really a switch station called by another name. Therefore, it is connected to the switch station bus. This connection is where the data input module receives its power and communication capabilities. The data input module can be mated to a specialty interface board. The specialty interface board provides eight sets of Form C dry contacts for indicating module input status. This interface board can be included inside of a data input module enclosure if specified at ordering time. A separate power supply input is needed for the specialty interface board.

The operating voltage of both data input modules is 12 to 20vdc and the amperage requirements are 10mA with no LEDs lit and 105mA with all LEDs lit. The number of inputs per module are eight and the voltage-out rating is 15 amps. Analog input drive requirements are 5vdc at 0.2mA. Additional input drive voltages are 12vdc at 3mA and 24vdc at 9mA. The rated minimum ambient operating temperature is 32 degrees Fahrenheit (0 degrees C) and the maximum rated temperature is 95 degrees Fahrenheit (35 degrees C). The dimensions are also the same as the other modules mentioned previously. The momentary data input module ordering number is 08-2290-02 and the maintained data input module ordering number is 08-2290-01.

The specifications for the specialty interface board are next. The operating power required is 350mA maximum and the output rating is

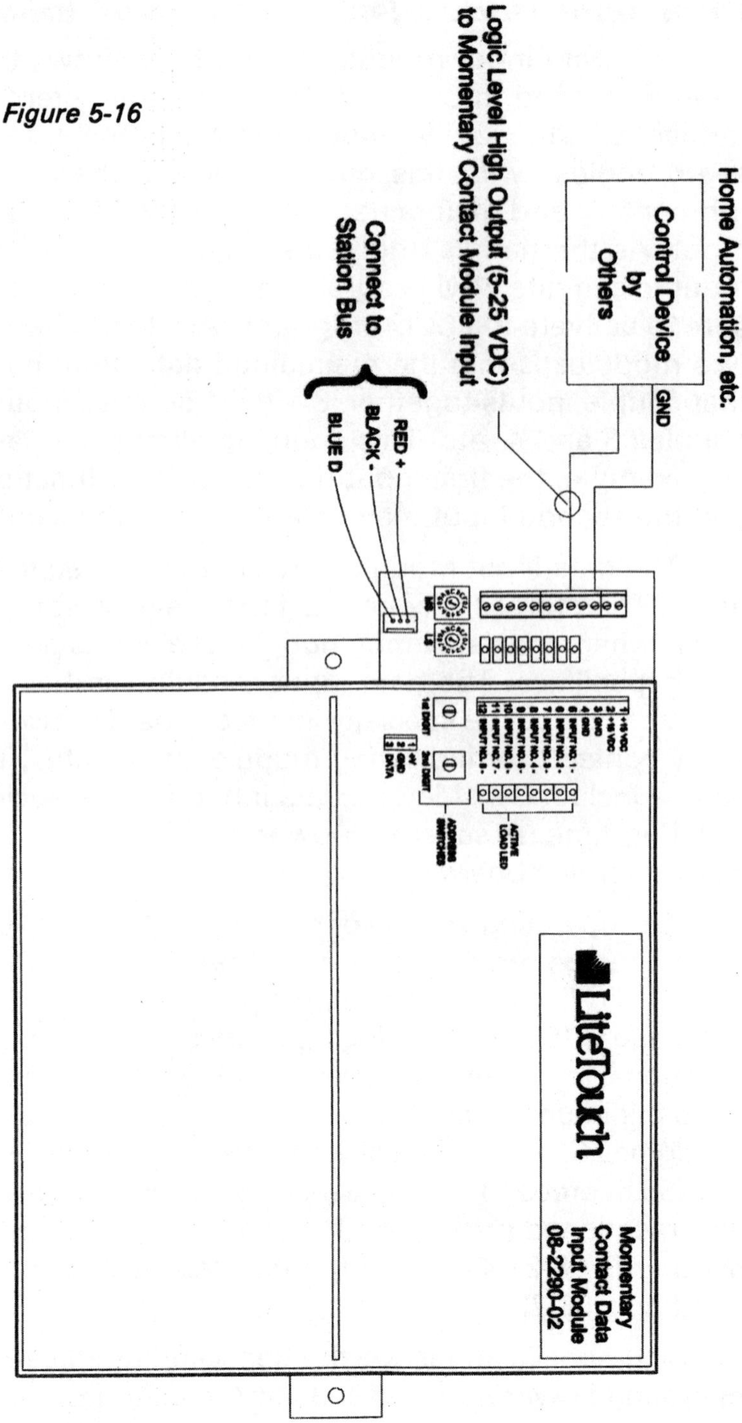
Home Automation, etc.
Control Device by Others
GND
Logic Level High Output (5-25 VDC) to Momentary Contact Module Input
Connect to Station Bus
RED +
BLACK -
BLUE D
LiteTouch
Momentary Contact Data Input Module 08-2290-02

*Figure 5-16*

1 amp at 24vdc. It has eight outputs and the dimensions are that of a standard control module (if it is ordered as a separate module and not included in the data input module can). This board can only be used in conjunction with a data input module. The ordering number is 08-2280-01.

### *Power Supply Module*

The power supply module depicted in Figure 5-17 is only one of two available. The second is called a dual power supply and consists of two single power supplies installed in one module can. The power supply provides dc power to the switch stations, the 2000 CCUs, and to many of the system control modules. All of the modules that need a power supply input draw about one-fifth of a single power supply module. One exception to this rule is the data input module, which draws the same as a switch station. Another is the analog output module. The analog output module uses about one-twentieth of a single power supply module. Power supply connections may be connected in parallel if more than one power supply module is located in a control module enclosure. LiteTouch recommends that not more than two power supplies should be paralleled into one switch station port on the back of the CCU or back plane board. LiteTouch also recommends that the power supply for the CCU should not be located more than 50 feet away from the CCU.

The output current capacity is 2000mA and the input voltage to the power supply should be 115vac (+10% at 30 watts). The external fuse is rated at 5 amps at 250 volts. The rated minimum ambient operating temperature is 32 degrees Fahrenheit (0 degrees C) and the maximum rated temperature is 95 degrees Fahrenheit (35 degrees C). Efficiency is rated at over 98 percent at full-rated load. The power supply dimensions are the same as the other LiteTouch control modules. The ordering number for the single power supply is 08-8610-01. The dual power supply ordering number is 08-8620-01.

## LiteTouch Scenario Controller

This stand-alone load controller has been around for quite a long time. Each scenario unit can control four separate dimming or switching loads. It mounts to a standard four-gang electrical box. In a basic installation, a 20-amp 120-volt feed runs directly to the four-gang box.

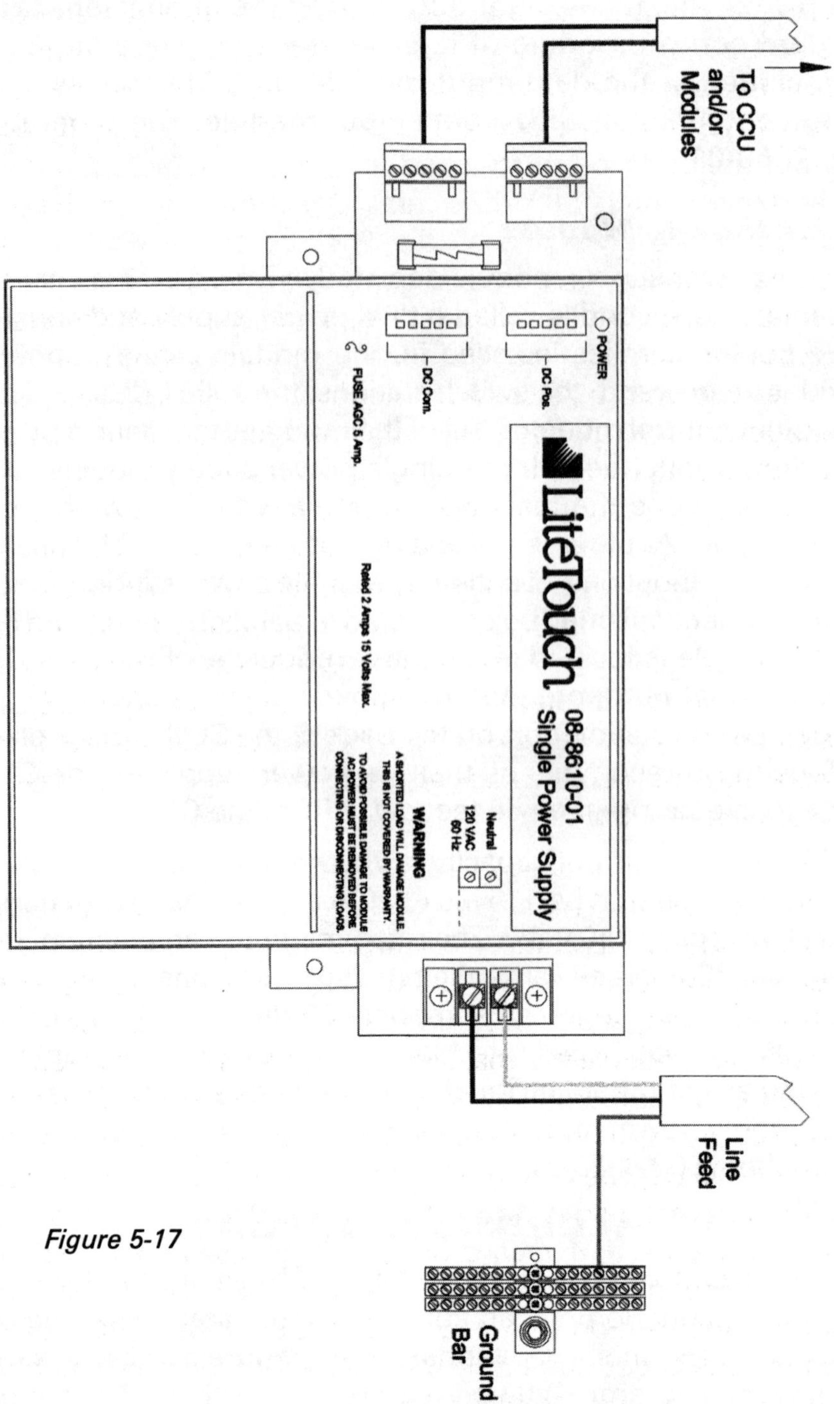
To CCU
and/or
Modules
POWER
DC Pos.
DC Com.
FUSE AGC 5 Amp.
LiteTouch
08-8610-01
Single Power Supply
Rated 2 Amps 15 Volts Max.
Neutral
120 VAC
60 Hz
WARNING
A SHORTED LOAD WILL DAMAGE MODULE
THIS IS NOT COVERED BY WARRANTY.
TO AVOID POSSIBLE DAMAGE TO MODULE
AC POWER MUST BE REMOVED BEFORE
CONNECTING OR DISCONNECTING LOADS.
Line
Feed
Ground
Bar

*Figure 5-17*

Then, the four high-voltage loads run to this same box. Remote switches can also be interfaced to the scenario controller. You can accomplish this task by running a three-wire low-voltage cable from the switch station box to the four-gang scenario box. A separate external power supply input needs to be introduced into this switch station cable at the plus and minus leads (normally this would be at the switch station box). A connector located on the back of the scenario unit allows the switch station cable to be attached for remote switching. This connector allows for the termination of the data lead and the plus/minus leads of the switch station cable. In addition, the connector permits an interface to other scenario units by attaching a cable to the control, data, and ground terminations (up to 16 scenarios may be interconnected). A quad dimmer module can also be integrated into the scenario system by the use of a special control module interface. Data inputs can also be integrated into the scenario system. Each scenario in the system must have a unique address if interfacing more than one unit. The factory default is set to zero and this is considered an address. The scenario is programmed from the front keypad by pressing certain function keys. Generally, this is a smaller control system than either the 2000 or 5000LC and it also tends to cost less. It has some interesting capabilities, so if you would like more information about the scenario system, contact LiteTouch technical support at 1-800-527-5527.

The power requirements for the scenario are 3 watts per unit and 2 watts per station. The supply voltage is 120vac and the supply current is 20 amps. LiteTouch recommends 12-gauge wire for the high-voltage line and load circuits. The unit rating is 2400 watts and the individual output rating is 800 watts (all four outputs not to exceed 2400 watts). The maximum module thermal dissipation rate is 40 watts (135 BTU/hour). The rated minimum ambient operating temperature is 32 degrees Fahrenheit (0 degrees C) and the maximum rated temperature is 95 degrees Fahrenheit (35 degrees C). The efficiency rating is over 98 percent at full load. The internal fuse is rated at 250mA/250 volts. The dimensions external to wall are 4 7/16 inches high by 8 inches wide by 7/32 of an inch deep. The order number for the scenario is 08-8200-XXX. The order number for the four-button remote switch station is 08-2804-XXX. The five-button remote order number is 08-2805-XXX. The order number for the nine-button remote is 08-2809-01.

## The CCU and Control Module Enclosures

The CCU has two different types of enclosures in which it can be housed. One is for surface mounting and the other is for both surface and recessed mounting. The Standard 2000, Compact, and the Elite CCUs will fit inside either enclosure. The surface-mount enclosure is smaller in size than the recessed type and should only be used for installations where the switch station and control module connections are kept to a minimum (like the book example install). The recessed enclosure should be used for the larger installations where the switch station count is 60 or more. This type of enclosure includes a hinge assembly that permits the CCU to swing out to access the connections in the rear. A wiring harness is provided with this enclosure to connect the CCU to the termination board. This enclosure also comes with removable transorb boards to protect the CCU from transient spikes. The order number for the surface-mount unit is 08-8025-01. The order number for the recessed unit is 08-8025-04.

The control module enclosures come in two different configurations—the four-module and the two-module enclosure. They both serve the same purpose and have the same features except that the four holds two more modules than the two. Both enclosures are available in either surface or recessed mount. The recessed enclosure includes a cover that overlaps the outside edge of the can by 1 inch all the way around. The surface-mount enclosure cover attaches to the can with the edges flush. The doors are always hinged on the left side of the can. Transorb boards are included with each enclosure and are mounted on the left-hand upper side. All control module enclosures are constructed of 16-gauge steel and are painted with a protective coating to resist corrosion.

The four-module enclosure dimensions are 37 inches high by 20 inches wide by 4 inches deep. The recessed lid for this enclosure measures 39 inches high by 22 inches wide. The surface lid measures 37 inches high by 20 inches wide. The two-module enclosure dimensions are 20 inches high by 20 inches wide by 4 inches deep. The recessed lid for this enclosure measures 22 inches square. The surface lid measures 20 inches square. The four-module recessed enclosure order number is

08-8024-02. The surface-mount four-module order number is 08-8024-01. The two-module recessed enclosure order number is 08-8024-06. The surface-mount two-module order number is 08-8024-05.

## Low-Voltage Cabling

The switch station control cabling is specified at 16-gauge, four-wire unshielded. The four conductors inside this cable are stranded and are color-coded red, black, blue, and white. The LiteTouch order number for this cable is 08-2103. A length amount will need to be specified when placing this order. The Belden equivalent of this cable is No. 8465. Only LiteTouch-approved switch station cables should be used with this system.

The control module communication cable is specified at 22-gauge, two-twisted pair shielded. The conductors inside this cable are stranded and are color-coded red, black, white, and green. Included with the foil wrap is a bare 22-gauge stranded conductor. Some LiteTouch equivalent communication cables do not have the same color codes mentioned in the specifications. The LiteTouch order number for this cable is 08-2105. A length amount will need to be specified when placing this order. The Belden equivalent of this cable is No. 9502. Only LiteTouch-approved communication cable should be used with this system.

## Transient Voltage Surge-Suppression

All electronic equipment is susceptible to transient voltage surges and the LiteTouch equipment is no exception. Direct lightning strikes are always a serious problem because protecting against them is nearly impossible. However, smaller surges can be controlled by installing the correct surge-suppression equipment. One study shows that about one-third of the transient spikes that enter the home are from outside sources. Power company switching and lightning are good examples of external transient spike sources. The remainder, it seems, come from within the home. Motor-type loads are usually the main culprits because of their "noisy" electrical operation. With all this potential surging activity, it is a good idea to protect your electronic equipment from its harmful effects.

LiteTouch recommends that at least a flat-blocker (no sine-wave tracking) type surge suppresser be installed at the main electrical panel (service entrance). This unit should be rated at 75,000 peak amps. They also recommend installing secondary surge suppressers rated at 50,000 peak amps at each electrical subpanel. This type of suppresser should have sine-wave tracking. If additional protection is desired, installing a unit rated at 90,000 peak-amps with sine-wave tracking for the main service entrance area is recommended. In this installation, secondary 50,000 peak-amp units with sine-wave tracking should also be installed at each subpanel. Extensive research has gone into finding ways to protect the LiteTouch system over the years. In addition to this research, input from field technicians and homeowners has also helped in this development.

This concludes the chapter on knowing your hardware. New equipment is always being developed in the field of home automation. Future improved versions of the hardware that you have read about in this chapter are inevitable. However, becoming familiar with the principles of the electronic equipment contained in this chapter will probably help when these changes or improvements are eventually made.

# 6

# Troubleshooting

This chapter concentrates on the Standard 2000 and 5000LC systems. However, most of the techniques covered also apply to the Compact and Elite systems as well. As every technician knows, problems are almost always unique to the specific job being worked on. There is constantly some new problem that has never been heard of before and, in all likelihood, some new way to deal with it. This chapter will try to cover as many of the known troubleshooting problems as possible by using LiteTouch recommendations and personal experiences.

Sometimes "gremlins" in the system can cause distress, but in the end, it always seems to get repaired. Many times these repairs are the result of a team effort between you, other technicians, and technical support. Imagination and good, sound troubleshooting techniques seem to be the keys to successful repairs. In troubleshooting, it is difficult to arrive at the conclusion that sometimes there is not a real problem. There may be times that a certain problem appears to be very serious. In reality, it frequently turns out to be something small like a programming error. Small problems usually act like large problems in this system.

In many instances, troubleshooting techniques are required to make a proper install of the system. This is especially true if you are not familiar with the installation that you need to complete. The rough installer may have forgotten some important aspects that you, as the finisher, will need to address. You will find that what is supposed to be, frequently is not, when it comes to a complex system like this. The installation will probably be the situation that tests your troubleshooting skills the most. This is not to say that you are running into trouble all of the time during your installs. It just means that when you do a considerable amount of home automation work, you will eventually come in contact with troubleshooting challenges. Troubleshooting during the installation period will be covered in this chapter.

There are a few important pieces of test equipment that you will need while troubleshooting this system. One is a good quality multimeter. Continuity testing is a very common procedure for checking switch station and control module cable runs. If you ever need to check continuity on actual electronic circuits, make sure you use the diode setting on your multimeter. Higher ohm settings may damage the internal components of the electronic equipment (also make sure the equipment is de-energized). Another item that is not always necessary but very helpful is a notebook or laptop computer. The Standard 2000 system CCU allows you to utilize diagnostic tests without a PC. However, the 5000LC and compact systems need a PC to accomplish proper troubleshooting tasks. There are always ways around not using a PC, but it may inhibit your troubleshooting progress.

The power bus monitor (Figure 6-1) is a very helpful tool for establishing communication faults. This unit can be supplied to you from LiteTouch. It plugs in to an empty data-out port on a control module. During the normal operation of the LiteTouch system (while the power bus monitor is plugged in), the green LED (1) should be on and the other LEDs should be off. When the CCU transmits to any control module on the line that you have plugged the monitor into, LEDs 1 and 2 will turn on for a short period. When the control module transmits a reply back, LEDs 3 and 4 will flash on for a short period. Each time the CCU transmits to the control module line, there should be a corresponding reply. The only time that the control module and the CCU do not communicate in this typical manner is on a power-up reset or when there is a power overload. If the green LED (1) is on normally, but you see transmissions appearing on LEDs 3 and 4 without transmissions on LEDs 1 and 2, check to see if the CCU is backed out to the first menu (normal operation menu—the one with the time and date). Reset transmissions are only answered in the normal operation menu. If the LEDs do not perform as they are supposed to, you have a wiring problem, a bad module, or the module does not have proper power.

The resident CCU tests and diagnostic software can be very helpful in pointing out a troubled area. These tests will generally indicate in an indirect manner what kind of problem the system is experiencing. The control station test, which is located in the system tests menu of the Standard 2000 CCU, is a good example of this. After this test has been

## Power Bus Monitor PC-2150
(Plugs onto an empty Data-Out port on a Control Module)

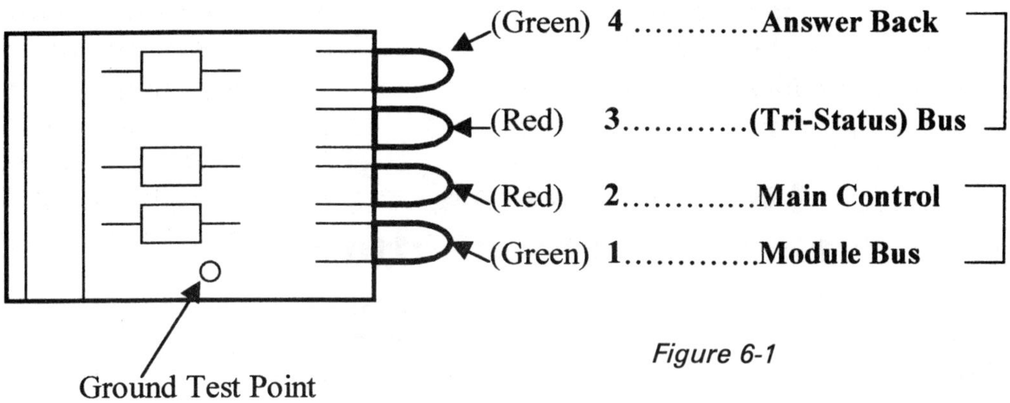


*Figure 6-1*

initialized, it may tell you that a station is not reporting with its LEDs on. This may indicate to you many potential problems. The switch station may be going bad or the control station voltage may be dropping too low because the LEDs are using up the available power. The switch station may, in fact, be going bad because the line voltage is at too low of a value. These problems may be caused by a faulty power supply or because the system was originally not designed with enough power supply units. This test result may also indicate that the switch station cable has too many switches installed on it. You can see the indirect manner in which the CCU tests can lead you in the right troubleshooting direction.

In this chapter there will be a section that addresses some of the real-life troubleshooting situations that a technician or electrician may experience while working with these systems. Another part of this chapter will deal with troubleshooting the 5000LC CCU. Excerpts from LiteTouch will be included to demonstrate how it would prefer the 5000LC to be tested.

Each technician has his or her own individual style when it comes to troubleshooting. When sharing their experiences with others, the knowledge that they impart helps to make all of us better at what we do. This chapter is an accumulation of many hard-learned troubleshooting experiences with which engineers, electricians, and technicians have had to deal. It is meant to reduce the amount of head scratching and frustration that goes along with learning something completely new. There is no way to cover every possible home automation troubleshooting situation in this book. However, continuing to use the process of independent thinking may help you to solve the troubleshooting situations that are unavoidably not mentioned in this text.

## Standard 2000 System Installation Troubleshooting

Many of the problems that will be discussed in this section will be transferable to the 5000LC system since it interfaces to much of the 2000 equipment. The Elite and Compact systems also use much of the same 2000 equipment and wiring.

A system that is completely installed by one individual can sometimes be a great advantage because that person knows exactly what has been previously accomplished. This type of situation is not always possible considering that some of the companies that install these systems may only do the rough-in work while completely different companies may end up doing the finish work. This will mean that a certain amount of investigation will have to be completed by the finisher if the house is to operate correctly. A meeting with the rough-in installer would be to the finisher's advantage, but this is not always possible or convenient. This section will discuss troubleshooting when the finisher is not furnished with all of the rough-in information.

Since most of the cabling is buried within the drywall, it will be a challenge to verify what the cables inside the CCU and control module enclosures are meant for (especially if they are not marked). This unmarked cabling problem is a very common occurrence, so continuity checks will be needed to establish exact cable-run locations. There is a wiring practice that consists of looping a switch control station cable from the CCU out to the switch stations then back to the CCU. This practice allows the station cable to be split into two runs at any switch-box location in the loop. This is a very good idea for balancing the switch load, but if the finish installers have not been informed of this, they may splice all of the switch station connections, thereby creating a perfect loop back to the CCU. If these two cables are then connected to the CCU, short-circuiting of the ports will occur. This can easily burn the ports out and damage the CCU to the point where it may not function at all. To avoid this, do a continuity check of all of the switch station cables. This should be done after the switch stations have been spliced, but before the switch stations have been installed (you will receive an erroneous reading through the station itself otherwise). Check between like colors and then between all of the wire combinations of every switch station cable. All of the switch station wires should show an open or infinite reading between them. If you find that there is continuity between any of the wires, you must find out where they are connected together and separate them before they are attached to the CCU switch station ports.

Another switch station issue is that sometimes the box itself will accidentally be covered over by drywall. This can cause a switch station-shorting problem inside the buried box if the cables are touching each other or something metal inside the box. It will also kill power and communications to the rest of the switch stations downline of the buried box. This situation can be avoided by checking the print to see if all of the switch station locations have been accounted for in the home. Any missing locations can be isolated to a certain area or wall location. Using a straight edge (like a level), you can check for wall bowing at the known switch height along the suspected wall. The level will stand off the wall more where a buried box is situated. Try to obtain an okay from the homeowner or contractor to cut into that part of the drywall where you suspect the box is. If you are wrong about the location and a hole is left in the drywall, the contractor or homeowner may be more understanding of this wall damage if you checked first. Make small cuts and try not to cut the hole any larger than the switch station plate will cover afterward. It is a slow process, but it will have to be accomplished sometime before the LiteTouch system is energized.

In rare instances, a switch station home-run cable (from CCU to the first switch station) is left out. Installing one after the home is complete can be a formidable task. You will most likely discover this after the system is powered up. However, it may be discovered during your switch station continuity checks. If you discover that a switch station home-run is missing, try to locate the best switch box to run the new cable to. This should be the one closest to the CCU and also with the best accessibility. In this situation, we will say that the best location is fairly near the CCU, but that you have to access the switch station box from the attic. In the attic over the box you have chosen to feed into, you may notice that a switch station cable is already running through the top plate. Fishing the new station cable through this existing hole may reduce the amount of time to complete the cable run and may also reduce the risk of running into a high-voltage cable if drilling is attempted. Once the cable is run from the CCU to the switch station box, you can splice it into the system (with power off).

Every once in a while you may discover switch station splices that are incorrect. In some cable assemblies the blue wire is colored a deeper shade and this can cause some splicing problems. The blue and black wire splices are sometimes swapped and this effectively shorts out the data line. Some LiteTouch pigtail assemblies (from the splice to the switch station) also have this darkened blue wire that sometimes is mistaken for a black wire. Bad lighting situations at the time the splice is being made can also account for this mistake. Check all splices for proper polarity before the switch station is installed.

Another situation that might arise is the possibility that too many switch stations are designated for a single cable run. LiteTouch recommends that not more than 20 switch stations should be installed per single cable run. This is a maximum amount, so single cable runs that accommodate fewer switches than this amount should be a common practice. You may discover this single cable run overage during the switch station continuity checks. If an overage is discovered, another switch station cable will need to run from the CCU to reduce the voltage drop that will occur along that single run.

Checking the control module communication cables is next. These cables are easy to identify and should run from the CCU to the transorb boards in the control module enclosures. These cables can also be daisy-chained from control module enclosure to control module enclosure once the original run is made from the CCU. There have been instances where these communication cables have been misused. They have been used as telephone cables and also have been inadvertently run to control station boxes. Completing continuity checks on these cables in the CCU and control module enclosures may be a good insurance policy (it also ensures that all of the module enclosures have communication cables running to them from the CCU). Try not to splice these communication cables together inside the CCU enclosure. If there are four cables, connect each cable to its own individual control module port.

The low-voltage power supply cables are identical in appearance to the control station cables; therefore, they can easily be marked incorrectly or run to a wrong location. Check continuity from the power supply cable inside the CCU enclosure to the power supply cable located in the control module enclosure. First, check to see if all of the wires inside the cable are reading open, then strip two of the wire ends in each enclosure (all power needs to be off). Wrap two of the ends together in one enclosure (make note of the two colors) and then take a reading from the corresponding wires in the other enclosure. You should have a short circuit if you are taking measurements off the correct cable. Untwist the shorted pair and then test to make sure that the reading is now at infinite or open. If all went well, you can connect the power supply leads to their proper connectors.

At the time of switch station installation, you may find that the one-gang boxes are too small internally to mount the switch station properly. This can be a big headache. The construction of the switch station will not allow any force to be placed on it. If the box is metal and you try to bend it to accommodate the switch, it will most likely misalign the mounting holes. In addition, the box will never bend enough to allow the switch to fit in properly. Plastic or phenolic boxes will only crack if pressure is applied to them. In this situation, box replacement seems the only alternative. It is very difficult to find a large enough cut-in type box to accommodate the switch station, so this is why it can be such a headache. When you come across this problem, you then understand why it is so important to make sure that the one-gang boxes you install on rough-in have ample room for the switch station.

After the Standard 2000 system is powered up and you complete your switch station test, you may find that too many switch stations are located on one port and power supply. For instance, you may find that three switch station runs go to one port and only one station run goes to another. At this point, you need to power down the system and separate the control station splices and then distribute them more evenly. Make sure that the switch station power supply allotment is also even. A single power supply for a few switch stations on one port while another port has a single power supply handling many switches is not right. Resplice until all control station ports are balanced with regard to power supply and switch station allotment. The power supply specifications allow for 35 control stations for a single power supply. This amount is really a maximum. If at all possible, keep the switch station amount per power supply lower than this.

You may notice while doing your switch station walk-around that some switches are not responding correctly. In fact, the engraved buttons don't even look like they correspond to the room you are in. Pull the station from the wall and double-check the address against the print. You may find that the switch station has been swapped from some other location. If a living room switch station button says "Exhaust Fan" on the front or "Bathroom Light," chances are the correct switch for the living room is in some bathroom somewhere. This incorrect switch location will still control the bathroom exhaust fan from the living room and the incorrect switch station in the bathroom will turn on the fireplace lights in the living room. These switch stations will need to be changed to their correct locations, however.

During the switch station walk-around you may also notice that when you push a button on a certain switch station nothing happens. You may try another button on the same switch station and still no loads respond. The button LED lights while you press it, but the LED does not stay lit once you let up on the button. Although this could be a bad switch, it is unlikely in this situation because the CCU port is still working and the dc voltages are good. Pull the station from the wall and check the address against the print. You will most likely find that the switch is not addressed correctly. Once the correct MS and LS ID address is dialed in, the switch will now respond to CCU programming. The reason the buttons would light only while you were pressing on them showed power was present to the switch (not necessarily data). The reason the LED would not stay lit once you pressed and released the button was due to the absence of any function/load data contained in the CCU. Once an address was dialed in that contained function and load data, the load then activated and the LED stayed lit. (Another term for this is an LED update.)

At this same station, you may press a button and the LED will light up, but no load will turn on. You notice this same situation on a couple of other switches also. You can go to the CCU and from the normal operation screen, press the Enter key. Then press the 1 key twice. Then press the Enter key again. The CCU LCD screen will display that the next station that is pressed four times will be identified. Go back to the trouble LED button and press it four times. The button will start pulsing to show you that the CCU has accepted the input. Go back to the CCU and press the 1 key to View and Modify Switch Assignment. The switch identity and the load(s) assigned will be displayed. Write down the load number. Press the Back key until the normal operation screen is displayed (the one with the date and time on it). Repeat this step for each button that the LED is lighting with no load responses evident. You will most likely see that the common element with every button is that it is supposed to control loads on one particular control module address. This points to a module problem. In this situation it could be many things such, as a breaker not being turned on, a bad module, or even a wiring problem. However, check the easiest and most obvious problem first. This is a brand-new installation and you probably made sure all of the breakers were on before you started the walk-around. This calls for a look inside the control module enclosure. Once you have opened the correct enclosure that houses the module in question, check for a power LED indicator on that module. We will say that it is lit. Now, check the MS and LS address dialed into the module. You will probably find that it has been dialed in wrong. Dial it in correctly and then check the buttons to see if the loads now turn on when you press them. This would not have been a problem to suspect in an established automation system. In this situation, if you were called in for a service call and the buttons were doing the exact same thing, you would probably suspect something else. Installation troubleshooting techniques can be quite different from standard troubleshooting techniques.

Let's use the same switch station for another example. This time the switch station does not turn on any of the loads and the LEDs do not stay lit when you press on a button. You know the switch station address, so you go to the CCU and press Enter from the main screen. Then press the 1 key for Examine/Update Switch Assignment. It will prompt you to input the first two digits of the control station. After you do this, press Enter. It will then prompt you to put the last digit of the address in and then press Enter. In this situation, the next screen will display the switch address that you have just entered and also that it is unassigned. Therefore, the reason the switch was not working was because the switch has not even been programmed yet. Select to assign the switch and then program all of the functions and load assignments for the entire switch station. Go back to the switch station and verify that all inputs are as they should be. Many times a switch station will somehow be left out during the programming stage, so this is common on a typical installation.

Oftentimes, during the finish stages of construction, custom furnishings such as built-in beds and china cases are not installed until the very end. This usually delays the installation of control stations since these are typically integrated into many furniture-type items. The control station wiring is usually lying on the floor or coiled up on the wall while awaiting the arrival of its final install area. At this point, the system is usually powered up and running, which means that these station cables are also live. This can be a problem due to the fact that someone else will probably have to handle the installation of the live control station cables into the new furniture. Special instructions need to be left with the contractor so that these cables are not shorted during final install. If it is possible, find the splice that leads to the coiled up cables and disable this part of the switch station circuit. When the furniture is finally installed, you can come back and reconnect the splice. This precaution may save some rather expensive CCU repairs later on.

High-voltage loads are another issue that can sometimes cause trouble during the initial installation. Using our original example switch station, we can show how a misplaced load can cause some difficulties. When you press one of the switch station buttons, nothing appears to happen. Then someone in the house asks rather loudly, "Who's playing with the lights?" The painter emerges from a dark hallway and claims that the lights keep going on and off. This should not be happening, but the button you were just pressing is turning the hallway light on and off. You programmed the house yourself and you are sure that you programmed the switch correctly. You then check the switch program for the trouble button at the CCU. The programming verifies that it will activate the correct load. Next, you open the control module enclosure and verify that the switch and CCU are activating the correct output. The correct red LED indicator on the control module is turning off and on with the button presses, so nothing seems amiss. However, further inspection shows that the high-voltage load is not connected to the correct module output. This is a small problem that is easily rectified. However, when loads are misplaced frequently throughout the control modules, it can get very confusing. Replacing the correct loads to the correct module outputs can sometimes take a long time to do, especially if the high-voltage loads are not marked correctly. This demonstrates how accurate the initial load placements need to be.

There will be many discoveries made during the installation period. You may find that the surge suppressors are not yet installed or that the security and telephone interfaces have not been connected. These are generally not serious troubleshooting-type situations, just areas that need special attention. Almost all of the troubleshooting areas mentioned in this installation section are only opportunities to fine-tune a system until the owner is satisfied. This is one of the great things about home automation—the ability to change much of what used to be unchangeable.

## Standard 2000 System Troubleshooting Illustrations

In this section, actual troubleshooting scenarios will be discussed. These troubleshooting events may assist you in your future dealings with the home automation environment. Overall, these systems are very sturdy. Like all complex electronic systems, however, they can experience difficulties. Lightning and surge damage will be included in this section. Lightning damage is a very challenging troubleshooting situation because of the random destruction that is usually incurred. Various other interface and user-type problems will also be addressed in this section in addition to general component failure troubleshooting.

## *Replacing a Bad Switch Station*

Upon arriving at a LiteTouch house for a service call, the owners tell you that they cannot control any of the lights on the second floor. In addition, the outside motorized storm shutters will not work from their master bedroom upstairs. They can control them over the telephone, however, and from a switch station downstairs. The first thing you will want to do is go upstairs and start pressing switch station buttons. You notice that the switch LEDs glow faintly while you press on them, but go immediately off when you let up on the button. Pull one of the trouble stations from the wall. Using your multimeter on the dc voltage setting, push the positive probe into the wire nut connection of the red wire (while the switch station is still connected to the pigtail). You then push the negative probe into the wire nut connection of the black wire. Note the voltage. It may read rather low, or it can read somewhere around 15 volts or so. Now, place the positive probe into the wire nut connection of the blue wire (leave the negative probe in the black wire connection). The voltage will probably be very low at this point. Maybe near 3.8 volts or so (since the line is being brought low). At this point, you probably have a bad switch station somewhere along this switch station cable run. Leaving the probes where they are, unclip the pigtail from the switch station (make sure that the switch station does not fall to the floor). Check back to your multimeter to see if the voltage between the data line and ground has increased. If not, replace the switch station pigtail connector and mount the station back into the wall.

Go to the next station that is not working and repeat these steps. Sooner or later when you pull a station from the connector, you will see a substantial rise in data voltage. When this happens, you have found the bad switch station. The rest of the switch stations will now work because the CCU has recognized that the line has good communications and will resume using that port. At the bad switch location (with the bad switch removed), check the voltage between the red and black wires again. It should read around 16 volts or better. Now check the voltage between the data (blue wire +) and the negative wire. The voltage drop between the positive and ground and between data and ground should be 1 volt or less. For example, your first positive and ground measurement is 17 volts. Your second measurement between data and ground is 16.5 volts. This is a sign of a healthy switch station line. There is no need to check for more faulty switches. If your second measurement was more than a volt difference (for example, switch positive and ground = 17vdc, and data and ground = 15vdc), then there probably is another switch station going bad somewhere. Continue to pull more switches until a healthy data voltage is attained. Generally, not more than one station goes bad at one time, unless lightning or other surges have entered the LiteTouch system. Lightning damage is known to destroy many switch stations, not to mention other system components. This may be a good time to check each system power supply to make sure that the all of the LED power indicators are lit. The status of these LEDs will indicate if any fuses are blown. Next, use your meter to check the dc voltages on the left side of the power supply. Verify the correct output.

The next item to examine is at the CCU. At the normal operation screen of the CCU, press Enter. In the right-hand upper corner, you will see a free memory percentage. If this value is less than 80%, you may want to repack the memory. Press the Back key to arrive at the main operation screen. Now, press the Option key. Packing the memory is accomplished by pressing the 4 key from this Options Menu. Warn the owners that the system will be sluggish during the memory pack and that it is best if they do not press any switch buttons. After the memory pack, make sure the normal operation screen is displayed before you leave the CCU. This troubleshooting illustration is complete except for one detail: you need to replace the switch station or send it to LiteTouch for repair. If you have a spare station, make sure that you address the new one exactly like the bad one you just removed from the system. When this switch station is replaced, check each switch button for proper program operation.

### *Lightning Damage*

The next illustration will address system lightning damage. You are called in to assess and repair a home with known lightning damage. The homeowners are distraught because most of their home automation system is not working. When you arrive, you make note that none of the switch stations has any power. The LEDs do not light at all. The CCU is displaying the normal operation screen, so at least part of the CCU seems to be operating. Next, you open the control module panels to inspect the modules. Many of the module power LEDs are not lit; however, all of the breakers are on. The power supply modules seemed to have been unaffected because the green power LEDs are all lit. Power is then shut off to all of the control modules except the power supply modules. You now remove the CCU from the enclosure to inspect the rear ports. Using your multimeter, establish that power is present at the three used ports, but no power is going out the switches. You disconnect one port at a time to see if the CCU will respond to the remaining ports, but nothing happens. You then disconnect all of the control station cables and reattach a small switch station pigtail to a switch station port connector. Reattach power to this port connector observing proper polarity for both the pigtail and power supply terminations. An operational switch station is then plugged into the pigtail connector. The modified port connector with the operational switch station is then plugged into one of the bad CCU ports. The switch station LEDs still do not light. You move the connector to each port and verify that the switch station does not operate on any of the three used ports. The fourth unused port, however, allows the switch LEDs to light. The CCU will have to be replaced since three ports are bad, but the fourth port may be good. This fourth port may be used to permit some lighting control until the replacement CCU arrives.

The next step is to establish what control stations are operational. Connect one control station cable at a time to a port connector, and then plug the connector into the good port. If any overheating occurs, remove the connector quickly and mark the problem control station cable for later inspection. However, in this case, you notice nothing unusual. You then measure (PS + and PS –) at the port and receive an 18-volt reading. You then measure (SW + and SW –) at the port and receive a 17.5-volt reading. Finally, you measure (Data and SW –) and receive a 17-volt reading. This switch station line appears to be good. You repeat these steps for each control station cable until all of them have been checked. In this installation, all control station lines check out, so it is safe to attach any one of them to the working CCU port. Start this station cable connecting process over, but this time do a control station test of each line by pressing the Option key of the CCU and then choosing 2-System Tests from the menu. From the system tests menu, choose Station Communication Test. Do this test for each control station cable. At the end of each test, write down all of the switch stations that have reported back to the CCU. This procedure will take some time because you can only use the one good CCU port to test each separate cable. Compile a list of good switches and compare it to the station install list if one is available. If a list is not available, you must physically check all of the switches for proper operation while you have each line connected to the good CCU port. In this testing situation, we will say that three stations did not report at all. You found the faulty station areas and verified that the switches were not operating. You then removed all three faulty switches from their locations. So far, the CCU and three switch stations need to be replaced.

Since the switch station port has limitations concerning power supply input and the switches are also limited by power supply input, only a select number of cables can be connected to the one good port. Ask the homeowners which switch stations they will require until the new replacement equipment arrives. Then, connect only these switch station cables to the good CCU port. This will allow a limited control of the house as long as you find that most of the control module loads are working properly.

You now return to the control module enclosures. Starting at each control module, you remove the fuses and check the continuity of each, replacing the burned out ones. Check for carbon arcing, melted, or burned PC boards. Make sure no outward signs of damage are evident. Verifying that the power is off, check to see if the loads from the house are shorted (you may need to call an electrician for this). When all loads have been verified to be short free, the next step can be completed.

The fuses have now been replaced and the modules do not appear to be outwardly damaged, so turn the power back on to each module. After each breaker is turned on, go to the module enclosure to see how the control modules are handling the introduction of power. In this troubleshooting case, all of the modules now have their power LEDs lit. You now turn each load on manually to verify that the module outputs still work (with the manual DIP switches located on the left side of the module). The breakers have not tripped and the corresponding module load output LED is illuminating. Check the actual loads by walking around the house and verifying load activation after each DIP switch is turned on. Now that all loads and modules are able to work in the manual mode, it is time to check the communications to each of these modules.

At the CCU, select System Tests from the options menu. Then select Module Communication Test. Let the test continue until every module has reported on every loop. Once this test is complete and it shows that all of the modules are reporting on every loop, we now know that all of the system control modules are both communicating and operational.

To check the control modules more thoroughly, plug the power bus monitor onto the data-out port on the last module of the communication line. Verify that the green #1 LED on the monitor is lit and that the other LEDs are off. Ask the homeowners to press a control station button while you watch the power bus monitor in the control module enclosure. Once they press a button, verify that both the #1 and #2 LEDs turn on for a short period (this is the CCU transmitting to the control modules). Then verify that LEDs #3 and #4 flash on for a short period (this is the control module transmitting back to the CCU). If the power bus monitor reacts correctly, then the control modules and the wiring to them are probably in pretty good shape.

It is now time to place the order for the new CCU and the three control stations. You may want to overnight this delivery if this equipment is not locally available to you. This way, the homeowners will only have to wait a minimal amount of time to have their system back to the way it was before the lightning damage occurred. (Another option is to place a "loaner" CCU in the home and send the bad CCU to LiteTouch for repairs.)

The last thing to do before the new equipment arrives is to download the existing CCU program into your PC. This will permit you to upload this program back into the new CCU when it arrives. Please reference back to the chapter on programming for this upload/download procedure. Once the download from the CCU is complete, name the file so that you will recognize it from the other LiteTouch files on your PC. It is also recommended that you complete a definition check and pack the memory while the file is resident in your PC.

Once the new equipment arrives, connect the switch station and power supply cables to the CCU connectors as they were before the lightning damage occurred (with the power off). Plug all of the control station and control module connectors back onto the new CCU and complete the upload (with the power on). Double-check the CCU power supply and switch station voltages, then secure the CCU into its enclosure. Address the new switch stations and install them in their proper location. Verify button and load activation on the new switch stations. Then, recheck the control module enclosures to make sure everything is operating properly. Sometimes lightning damage can show itself at some later time, so this is a good precaution. You may want to check to see if all of the surge protectors are still showing their operating LEDs. Some suppressors may have internal fusing or may even need to be replaced after a lightning strike. If the CCU is using its telephone interface or modem, the telephone system may be another area to examine. Once you have checked the entire house again and feel that it is completely back to normal, the job should be totally complete.

Although LiteTouch usually undertakes equipment repair, it would not hurt to discover some of the discreet components that are harmed by lightning damage. Lightning-damaged electronics typically have no set pattern and, for the most part, is rather random. Of course, all integrated circuit chips are susceptible, but any type of communication chip on the printed circuit board is typically a very good target for lightning damage. Some of these communication chips are located fairly close to the data ports on the PC board of the control module, so it is a high probability that surges may enter the system through the communication cabling.

The switch station printed circuit board damage can be somewhat random. The switch station board transistors and IC chip can be common targets. Sometimes more hearty discreet components like resistors are destroyed. This kind of damage may be secondary to another board component failing due to the lightning surge. The Standard 2000 CCU is an industrial-type controller and is very complex. The easiest component that I have seen affected is a large ceramic resistor. It was melted out of one of the internal boards and when it was resoldered back in, the CCU worked perfectly. This is hardly any helpful experience concerning lightning damage on the CCU. LiteTouch has the real experience needed to make successful repairs to the CCU and all of the components that it markets.

### ***Frozen Relay Module Contacts***

The next troubleshooting case involves a load that will not turn off. A switch station in the kitchen of a custom home controls a large exhaust fan. When you walk into the kitchen, you can hear the constant droning of the exhaust fan blower motor. The owner claims that this blower motor would every once in awhile not shut off when he pressed the corresponding switch station button. "It was a nuisance up until this point," he says, "but now it stays on all of the time." You inspect the button and the LED indicator on the switch button is not lit. You press the button, and the LED lights, but there is no corresponding change in the blower motor. You then access the CCU and press Enter from the normal operation screen, then press the 1 key twice, then you press Enter again. The CCU display tells you that the next button pressed four times will be identified. Go to the problem button and press it four times. The LED will pulse to tell you that the CCU has accepted the input. Go back to the CCU and the display will give you the option to View and Modify Switch Assignment. Choose this option and the switch address plus the function/load information will be displayed. In this display screen, it tells you that the assigned load is 32-4 and that the switch function is a toggle. You then locate the enclosure that contains the control module 32-4. This module turns out to be a relay-type and after observation, you find that the CCU is, in fact, activating the correct output. However, the relay output is frozen in the on position. The contacts are obviously welded together. You then check the nameplate rating on the blower motor and find that it is rated very near the contact limitation for the relay module. The blower motor will draw more power on startup than the displayed amperage on the nameplate; therefore the contacts are sticking due to this initial over-amperage start-up. There is an extra relay module output that you can connect the blower motor to. Sooner or later, however, it will fail this new set of contacts. The best repair in this situation is to install an auxiliary relay that is slaved from the relay module. Since the old relay output is frozen, you will want to use the extra relay output that you discovered earlier. This will call for a program change at the CCU, so that the button in the kitchen will instruct the CCU where the new load is located. If you are not an electrician, you may want to call one to install the auxiliary relay off the new LiteTouch relay output. The new module relay output will only control the coil part of the auxiliary relay. A separate power feed will need to run to the controlling contacts of the auxiliary relay so that the blower fan will receive proper power. The auxiliary relay contacts should be rated well above the blower motor nameplate rating.

After the auxiliary relay is installed, the switch button programming needs to be changed. You can do a load assignment check to see if any other switches in the home control this blower motor, but this is highly unlikely. Barring any more switch stations in the kitchen that may control the fan, this is not the type of load that is usually controlled from any other location. Enter the switch station number into the CCU and select the Change Assigned Load from the list. Now add the new load address that slaves the auxiliary relay to the blower. After this has been completed, back out to the normal operation screen and test the kitchen button for proper blower activation.

### *Random Load Cycling*

On the next troubleshooting illustration, the homeowners are having a problem with random load activations. They complain that many of the loads in the house are either turning off and on or dimming up and down while no one is pressing any switch buttons. If this effect was concerning one load or a certain set of loads, it may be a switch station button problem. (The older S-Series switches were more prone to button sticking.) It could still be a switch station-type problem, but after talking more with the homeowners, it seems to be an entire systemwide problem. Most likely, the CCU has incurred some transient spike and the logic states are a little confused. From the normal operation screen of the CCU, press Option. Then, select 5-Diagnostics from this menu. The diagnostic menu will then appear. It may be helpful to initially select the first option 1-Run Switch/Load Definition Check. After this test is complete and no errors are found, select the third option 3-Clear RAM and reset CCU. You may want to inform the homeowners before you start this procedure because it will reset the CCU to some of the factory defaults. The switch/load information will still be present, but all of the loads that are currently running will be shut off. This could be alarming to them, especially at night. The CCU will then clear RAM and reboot itself. If this procedure does not stop the random cycling of loads, you may try initializing the CCU.

## Initializing the CCU

The following is an excerpt from LiteTouch for initializing the CCU. This procedure will result in the total loss of programming information in the CCU. It will wipe out all of the switch/load, time flag, and user parameter information. A download of the CCU is recommended before this procedure is started. If the current programming information is corrupted, however, you may not want to reuse this file to upload back into the CCU. It would be an advantage at this point to have the original upload file for the CCU on your PC. This use of the original file will probably be a better bet than using the current download file. The only other option is to program the CCU from scratch. If you have any trouble making a decision on what exactly to do, call LiteTouch support for advice. The source for the following information is from LiteTouch and will initialize the CCU EEPROM. This EEPROM area is divided into three separate areas: Switch and Load area, Time Flag area, and User Parameter area (this is where the astronomical clock information is stored).

## Standard CCU EEPROM Memory Initialization

Press [Enter] to get to the Main Menu.

Press [5], the CCU will prompt you for an access code.

Press [9 2 4 Enter] the CCU will prompt you for the Number of desired test or action.

Press [1], the CCU will display: examine_eeprom.

Press [C], the CCU will display: INITIALIZE TIME FLAG AREA

are you sure? (1-Y/2-N)

Press [1], the CCU will initialize the time flag area, and when it is finished, it will add to the display:

exit = 1, more commands = 2

Press [2], the CCU will display: examine_eeprom.

Press [D], the CCU will display: INITIALIZE SWITCH and LOAD DEFINITIONS

are you sure? (1-Y/2-N)

Press [1], the CCU will display: This will take up to 60 seconds, when it is finished, it will add to the display:

exit = 1, more commands = 2

Press [2], the CCU will display: examine_eeprom.

Press [E], the CCU will display: INITIALIZE USER PARAMETERS in EEPROM

are you sure? (1-Y/2-N)

Press [1], the CCU will initialize the user parameter area and when finished, it will add to the display:

exit = 1, more commands = 2

Press [2], the CCU will return to the Main Menu.

*WARNING: Initializing an area means the loss of all information in that area. That information will have to be reprogrammed (Source: LiteTouch).*

After completing this initialization supplied by LiteTouch, reprogram the CCU. If the CCU still continues to turn loads on and off randomly, call LiteTouch to see if replacing the CCU is in order.

### Relay Interface Trouble

In a particular LiteTouch installation, a series of motorized storm shutters were interfaced to standard six-channel relay modules. This arrangement worked well for some time until the shutter motors began burning out. Whenever the relays were tested, they performed flawlessly. For a time, the shutter manufacturer was blamed for substandard shutters. Another manufacturer installed new shutters, but these shutter motors also began burning out. It was decided that by using the standard relay modules, a power conflict to the shutter motors had been arising. This was not a situation that could be proven beyond a doubt, but it was a definite possibility. Interlock relay cards were then ordered to retrofit the standard six-channel relay modules. These cards replaced the standard relay cards located on the right high-voltage side of the relay module. The motor leads from the shutters were connected to these new interlock cards in such a way that a power conflict would not be possible (proper motor-lead pairing to certain interlock output numbers). This interlock installation is still working flawlessly and no more shutter motors have been damaged. The newer eight-channel relay modules come with interlock capabilities also.

### Switch Station Wiring Failure

In one installation, a CCU switch station port became nonfunctional because of a short somewhere along the switch station cable run. This could have easily been a switch station failure or the switch cabling causing this kind of problem. On the nonfunctioning CCU port, the connector was removed. Each control station cable connected to that port was then isolated from each other. The individual cables were tested, but only one cable indicated the short that was making the port go nonfunctional. The remaining good cables were then reattached to the port that was now functioning. It was then a matter of inspecting all of the switch stations that were not working on the one cable run. The problem was broken down to one line containing 10 switches. Every nonworking station was pulled away from its box and each switch and connection was checked. After many switches had been removed, a blue (data) wire was discovered to be shorting against the switch station's back heat-sink plate. This data wire had accidentally been stripped back too far, permitting about a half-inch of bare conductor to come in contact with the back of the switch. Properly resplicing the connection solved the entire nonfunctioning port problem. The switch station cable was then reattached to the proper CCU port.

### Switch Station Failing to New Address

Although switch failures can be a common problem in this system, sometimes they are not as easy to isolate as the earlier example demonstrated. In one installation, a homeowner complained that a bathroom light would always be turned on. He said that he distinctly remembered turning it off many times. When he returned later, the light would be turned back on. No one else had been in the house, so there was no way someone could be turning it on. The control modules were tested, as were the load assignments. The only point of control for the bathroom light was the switch station address inside the bathroom. This was indeed a head-scratching-type problem. Probably in an act of desperation, a switch communications test was performed. The test stated that a station had reported for a second time. It was deduced that the owner was probably pressing a button on a duplicated switch somewhere else in the house, which was turning on the bathroom light. There are many ways to locate this duplicate switch, such as isolating the port of the true address switch (by taking the correct switch station out of the bathroom). Then, you can isolate the port of the erroneous address switch by conducting another switch communications test. The switch addresses on the port with the erroneous address can then be written down from the list that appears on the CCU.

Walk through the house and locate the switch stations that you have on the list. The erroneous switch station will probably be somewhere near these listed switches. You can also push the buttons on each of these switches to see if they, in fact, control the proper loads or act unprogrammed (LEDs will not stay lit when pressed). The tip-off of an unprogrammed switch button is that the erroneous address switch will most likely not be configured like the true address switch. This means that even though the address is the same for both, they probably will not have the same amount of buttons on each switch station. When you suspect a switch, go to the CCU and press Enter. Then press the 1 key twice and press Enter again. Go back to the suspected switch and press one of the buttons four times. Go back to the CCU to read the address from the display. It will probably be the duplicate bathroom switch address. Remove the switch and check the address dialed in. It will most likely be something totally different than what the station is actually reporting back to the CCU. Replace the bad switch station and make sure that the proper address is dialed into the new one.

### *Control Module Communication Failure*

When a control module fails totally, it generally stops any further CCU communication down-line of it. This is because the communication jumpers are wired in series from module to module inside their single enclosure. Usually, this does not affect any of the adjoining module enclosures because they are generally wired in parallel with each other. For example, separate communication cables are commonly run from the CCU to each module enclosure and then to the transorb board inside (sometimes they are daisy-chained in parallel). Inside the four-module enclosure, the first module receives communications from the transorb board. After that, a jumper is run from the data-out port of the first module to the data-in port of the second. This continues until all four modules are connected to the communications bus. At the last module in the enclosure, the data-out port has nothing connected to it. If the first module failed, or had no power supplied to it, the second, third, and fourth modules could not communicate with the CCU. Therefore, a problem with the first module can make it look like there are more problems with the remaining modules in the enclosure. In most cases, the rest of the modules in the enclosure are in good operating condition. The status of the remaining modules can be tested by removing the jumper from the data-in port of the first nonworking module and then plugging that jumper into the data-in port of the next module down.

Communications should now be restored to the remainder of the enclosure modules. Use the power bus monitor clip to verify this afterward. This moving of jumpers can be helpful when troubleshooting communication problems with control modules. It also helps to keep most of the home automation system running in the event that there is not an immediate replacement for the bad module. All of the fuses and circuit breakers should be checked prior to purchasing a new control module or sending one in for repair.

### *Power Supply Problems*

The LiteTouch power supply module has been reengineered a few times over the years. You can recognize the older power supply units by their maroon color scheme and the many low-voltage wires coming off of the supply connectors. You may see some of the outputs of these older modules with an external 3300 microfarad capacitor connected across the positive and negative for power smoothing. Many output variables have been improved in the newer power supply units. They have always been known to be very sturdy and reliable. One of the problems that you might find in certain existing systems is that there may simply not be enough of them installed in a system to keep the electronics healthy. As a technician, you can spot this situation by observing lower-than-normal switch station voltages or too many control modules being powered by too few power supply modules. If open module spaces in the enclosures are available for adding an additional single or dual power supply, it may be wise to make the addition. Extra high-voltage input circuits and output cabling can be added to complete the additional power supply installation.

The power supply operating LEDs should always be checked whenever you are on a service call to a home. Many times, the reason for equipment failure is simply due to the fuses being blown to the power supply(s). A power anemic system should always be suspected when a particular system's equipment continues to fail on a regular basis.

### Switch Station and Control Module Cabling

Earlier, there was mention of a use for the fourth white wire in the switch station cabling assembly. This fourth white wire can be used to boost the capabilities of the existing switch station wires. It is recommended that when voltage drop on the switch station cabling system has become a problem, the fourth white wire may be integrated into the negative power component. This connection of an extra negative conductor into the switch system usually begins at the CCU. The white wire is connected to the black (negative) wire of the switch station cable (SW –). This white wire is then connected together at each switch-station box until the box that is farthest from the CCU is arrived at. The white (negative) wire is then introduced back into the negative wiring system by connecting it to the black (negative) wire of the farthest station. Check continuity between the black wire and the white wire before you make this connection. You should read a low-ohm closed circuit. If you do not read a closed circuit, there is a switch-station box somewhere that does not have the white conductors terminated together.

There have been instances where the extra white wire of the switch station cabling has been used to bring (SW +) power to the farthest station. This was accomplished in the same manner as mentioned before, only this time the white wire was attached to the red (positive) wire of the switch station cable at the CCU. Then, the white (positive) wire was reintroduced back into the red (positive) wire at the farthest switch station. This would have to be accomplished with the power to the switch station turned off to avoid shorting. In addition, the continuity check should be accomplished with the power turned off. LiteTouch recommends using the white wire for the negative component (rather than using it for the positive component).

Some of the equivalent communication cable that is specified for the control modules is not color coded in the same fashion as mentioned earlier in the book (not the same color codes as the LiteTouch cabling). This can cause some communication wiring identification problems. Many times, colors like black are used more than once inside the cable assembly. These black wires are usually twisted in a special pair and many feel that this is a sufficient enough identification to attach them to the control module bus without further verification. While this may be true, a continuity check of these like-colored conductors would give the technician an extra confirmation.

### The Standard 2000 CCU

As mentioned earlier, the Standard 2000 CCU is a complex industrial-type controller, and it is difficult to troubleshoot the internal electronics. However, there are many ways to spot trouble and also a few troubleshooting techniques that can be implemented to keep the CCU performing properly.

Every so often, the CCU will act strangely or lock up. You may observe sometimes that all of the character lines on the LCD screen are darkened upon power up or that it stops responding to inputs. A power reset of the CCU will sometimes clear these anomalies.

On one occasion, a certain CCU needed to be programmed. Most of the internal operations of this CCU functioned properly, except that its LCD screen was totally blank. Normally, you could program from the PC and upload the program into the CCU at this point, but the RS232 port was also nonoperational. This unit was eventually sent back to the factory and later replaced. For the moment, the homeowner needed some control of the system. Therefore, an attempt was made to program the CCU blindly from the front keypad. This blind programming was accomplished at the CCU by using previous memories of what the screen should display with every input. When you reached a point where you could not visualize what the screen should say or became lost in the menus, you just backed out until you knew you were at the normal operation screen again. From the normal operation screen, you could then visualize your way through the menus once again. This kind of situation is not common and no one usually expects you to program under these conditions. However, you never can tell.

When a CCU software version is older, the homeowner may opt for an upgrade. Before an upgrade is started, it is recommended that you download the program file from the CCU first. Changing the software version is accomplished by accessing the inside of the CCU. Keyed machine screws are removed to allow the front cover to come off. Carefully unplug the keyboard ribbon cable from the printed circuit board, noting the polarity of the keyboard connector. There should be a grounding jumper from the front panel to the rear that also needs to be removed. Locate the older EEPROM chips marked A3 and A4. Remove both of these chips with an IC extractor, making sure that the pins are not bent. You may want to use a grounding wrist strap so that these static-sensitive chips are not damaged. Then, carefully install the new software version chips in the A3 and A4 sockets. These need to be installed in the proper direction into the socket. Again, make sure that the IC pins do not receive any damage. On some of the older CCUs, there is a piggyback voice chip installed for the telephone interface. On the newer CCUs, a separate socket for this chip is provided. Reassemble the CCU. The software version is now updated. To check the CCU firmware version, press the 8 key from the Main Menu. Pressing the 8 key will also access the CCU serial number and module code version.

### *Dimming Loads*

Troubleshooting dimmer module problems is easier once you acquire a little bit of knowledge about how they work. For example, a dimmer module will output floating voltages when all of the loads on the module are off. Once one of the loads is turned on, this floating voltage should disappear from the remaining dimmer outputs. The reason for the floating voltages when the dimmer module loads are off is because of triac current leakage. It is not unusual for this floating voltage to be higher than the input voltage to the module (sometimes around 138 vac). This floating voltage will not have the power to turn on a load, but it can confuse you when you are measuring output voltages in a troubleshooting situation. The reason that these floating voltages should disappear after one of the loads is turned on is because of an air gap relay that is integrated into the dimmer module. This air gap relay is only activated when there is an actual load on the module (power being used). It is important to note that touching an off dimmer output can give you a nasty shock, so treat these dimmer loads as though they are always hot (even if one load on the module is on).

The dimmer module is extremely susceptible to any type of shorting situation. These modules can be damaged from either a short to ground (to grounded metal) or a short between leads (like a transformer burning up). The dimmer module outputs commonly fail in the "on" state, but can fail in the "off" state, too. These outputs are isolated from each other inside the module. Therefore, a short on one output does not usually affect the other outputs. Sometimes the output's serviceable fuse will blow before damage is done to the dimmer output, but this is not something to count on. All high-voltage loads should be checked thoroughly before they are attached to the dimmer module.

In the CCU programming environment, a minimum turn-on level can be set anywhere from zero to 50 percent. The factory defaults are set at 25 percent. These settings determine when the dimmer module will send its output to the load. The dimmer module output may be "on" to a 20 percent level internally, but if the minimum turn-on level at the CCU is at the default setting of 25 percent, the output will not activate until five more percentage points are reached. This feature can eliminate the dimming dead zone that most dimmers display.

This minimum turn-on level was used in an actual troubleshooting case. A homeowner was complaining that a fixture in her bathroom was cycling on and off at a very low dimming level. To try to solve this problem, the CCU RAM was cleared. In addition, a complete CCU initialization was accomplished, but the problem persisted. It appeared to be a small problem and it did not seem realistic to take drastic steps like replacing all of the equipment that could possibly be causing this to happen. The bathroom light seemed to be oscillating at a percentage level below 20 percent, so a troubleshooting hunch was tried. The CCU was accessed and the minimum turn-on level was found to be set to 10 percent. This minimum level was then reprogrammed to a value of 25 percent. After this change, the bathroom light did not oscillate at the low-percentage level anymore. This repair may have disguised another system problem. However, in the event that the system does display this problem again and this time, maybe on a wider scale, more drastic measures can be taken then. The drastic measures that may need to be taken might cost a considerable amount of money. The homeowners will probably be more agreeable to spending the money if the problem is on a much larger scale.

There may be times that a troubleshooting case is solved with only a slight programming change. For example, one homeowner said that she was having a dimming-type problem in her home. One group of lights in her bathroom had one switch button (with the function of a dimmer) that controlled the complete set. She would set the dimming load to the desired level by holding the button until it was reached. Typically, someone else would then change it to some different level. She wanted something done about it. This programming change was not difficult to figure out once you discovered what she really wanted. The button's function was changed to a scene preset with only the one load assigned to it. The level was then set to 35 percent. She tested the button and said it was almost what she wanted, only she preferred it to be a little less bright. The load level was then changed to 30 percent and this setting was perfect. Now, no one will change this set level because every time someone turns that button on, it will dim to the desired 30 percent level (unchangeable at the station while the system is locked).

### *Equipment Interfacing Problems*

The exact interface needed for a piece of equipment varies with each product and manufacturer. Generally, nonsystem hardware will send an output to the LiteTouch data input module (or station) for home automation control. An output from LiteTouch to nonsystem hardware can come from a variety of sources including modules like the low-voltage, dry-contact relay. Of course, there is other interface equipment that intelligently mates the LiteTouch system with the nonsystem equipment.

Once home automation interfaces are established and the systems are up and running, troubleshooting becomes a little trickier. This question always arises: Is it the LiteTouch system causing the problem or the nonsystem equipment? This question can go back and forth, especially if the contractor of the nonsystem equipment does not understand the interface or the LiteTouch system. Many times the LiteTouch technician has to understand the nonsystem equipment as well as his own equipment so that an accurate troubleshooting decision can be made. This quality tends to make the LiteTouch technician very valuable to the homeowner.

Interfacing can be complex, but two troubleshooting rules to follow are:

1) If the proper input is arriving from the nonsystem equipment, most likely the problem is with the home automation system.
2) If the output is leaving the LiteTouch system properly, most likely the problem is with the nonsystem equipment.

If only troubleshooting interface problems were as easy as these two statements sound. Unfortunately, usually they are not.

The future of home automation interfacing will most likely include a set of standards. At times, the nonsystem equipment that LiteTouch will operate (or that will operate LiteTouch) really has no set of interfacing standards. The interfacing compatibility issue is commonly addressed in the planning stages of the LiteTouch installation. For example, a spa company representative will sit down with the LiteTouch designer and discuss interfacing capabilities. The spa company will list the equipment and specifications that will be used in the home and the LiteTouch designer will create an appropriate system design to accommodate the features and specifications of that system. This cooperation will eventually produce a compatible interfacing system that both can agree on. The technicians from the spa company and the technicians who install the LiteTouch system then make the interface a reality. The day will arrive where home automation interfacing will be as common as the mating of a television set with a VCR. Standards will be set and all equipment manufacturers will comply with these standards, much like the television manufacturers do today. Until then, it is up to the designers and technicians to work out each specific equipment interface situation. Currently, there are some home automation interfacing standards set in place by certain HVAC, entertainment, and other peripheral equipment-type companies. So it is slowly becoming a reality.

## 5000LC Troubleshooting

The 5000LC has some troubleshooting advantages over the Standard 2000, since each system card can be removed to identify problem areas. The back plane board can be tested separately from the system boards; therefore, the complete 5000LC disassembly permits troubleshooting isolation. As you know, problem isolation makes troubleshooting much easier.

The monitor card LEDs indicate power status and station/module bus communications. The C2000, CPU and modem cards have a seven-segment LED display that shows proper card and system operation. In addition, a Windows-based diagnostics program named CP5000 can be run to conduct further troubleshooting activities. Between all of the included diagnostic features, the 5000LC system can make troubleshooting much easier. The following are the overall guidelines supplied by LiteTouch for troubleshooting the 5000LC:

1) Check wiring. Are the correct polarities and wiring colors observed?
2) Use a voltmeter to ensure proper voltages are obtained from the Power Supply Board and the Station Bus Power Supplies.
3) Are components on any board getting too hot to touch? A short could be indicated.
4) Are any fuses blown?
5) Start with the Monitor Card and add the other cards one by one, and note when the problem occurs.
6) Try removing individual station ports or module ports to see if the system is being loaded down.
7) Do you have the latest firmware in the cards and the latest software on the computer?
8) Are you using a null-modem or a straight-thru RS 232 cable? Only straight-thru cables will work.
9) What changed since the last time the system was running? Did lightning strike or did the power go out?
10) Does turning on and off the controller fix the problem?
11) What happens when a station or module diagnostic is run? 6-channel relays and dimmers give phantom bad messages on the next address up if none is present.
12) Did the job programming get uploaded correctly? If so, how do you know? (Source: LiteTouch)

Each of the next sections will deal with the individual components of the 5000LC system.

### Back Plane Board Troubleshooting (Station Bus)

The back plane board troubleshooting consists of three areas: station bus, module bus, and power supply. The three-row DIN 64-pin card connectors are also located on the back plane board. The station bus troubleshooting will be discussed first. You may want to reference back to the figures in the install area of the 5000LC while you review these sections.

Station bus problems are sometimes detected by taking voltage measurements, much like the Standard 2000 system. Voltage measurements should be taken at each station port. They are identical, so an individual port will be used for the example. Starting at the (PS +) and (PS –) terminals, measure the dc voltage. Depending on the particular power supply or power supplies used, this measurement should read from 15 to 24 volts. If this voltage is lower, remove the three wires from the switch station (SW +, SW –, and Data) terminals to see if the voltage rises. If the voltage rises substantially, there may be a bad switch on the cable run. If it rises a few volts, too many switch stations may be installed on the line. Replace the (SW+, SW –, and Data) cable onto the port terminals. Now, measure the dc voltage to the (SW +) and (SW –) terminals. It should be within a half-volt of the (PS +) and (PS –) readings with no switch load. It should be within 1 volt with a switch load. If this voltage is not within these parameters, the port may be overloaded. Next, with no switch button presses occurring, measure the dc voltage from the (SW –) and (Data) terminals. This should be within a half-volt of the (SW +) and (SW –) reading with no load. It should be within 1 volt of the (SW +) and (SW –) readings with a load. If this voltage measurement is extremely low, a bad switch station may be located on that switch station port. If this is the case, remove the switch stations one at a time (as was explained in the Standard 2000 troubleshooting section) until the voltage rises back to normal. You may want to isolate the exact cable run that is giving the trouble (if the station port has more than one station cable running to it). This cable isolation procedure may cut down on the search time for locating the faulty switch station. The following note is from a LiteTouch troubleshooting excerpt:

Note: If at least one station port has power, the DATA lines on the other station ports are pulled high to prevent spurious communication on nonactive ports (Source: LiteTouch).

The back plane board includes transorb protection for the system cards. If these surge protection circuits are damaged, the back plane board will have to be repaired.

### Back Plane Board Troubleshooting (Module Bus)

The control module bus is located in the upper section of the back plane board. This board only routes the control module communications to the Monitor system card. If a communication problem arises, look to the monitor card for troubleshooting capabilities. The monitor card will have its own troubleshooting section, so please reference this area for control module communication problems.

### Back Plane Board Troubleshooting (Power Supply)

The power supply is located on the right, high-voltage side of the back box partition. It supplies the 5 volts positive dc to the system cards. The following is a note provided by LiteTouch regarding the high-voltage input grounding: The incoming ground must be wired into the Ilsco ground lug (as indicated by the ground symbol) to meet UL regulations.

With the power supply energized, make sure the power LED is illuminated. If this LED is not lit, turn off the power and check the board fuses for continuity. Replace them if they are blown. If replacing the fuses does not make the LED light, the LED or the power supply itself may be defective. Measure the voltage on J1. If the voltage is 5 volts dc (plus or minus 0.15), then the power supply is okay. If the voltage is not within these parameters, it may be due to excessive power supply loading. Power supply loading can be checked by removing the four-pin cable that leads to the back plane board and then completing the voltage measurements again. If the voltage is still not within the accepted parameters, the power supply is most likely bad and must be repaired or replaced.

### Monitor Card Troubleshooting

The monitor card is always installed in the right-hand slot of the back plane board. As mentioned earlier, it controls the power to the other system cards and indicates the module and station bus activity. As with all of the 5000LC system cards, the monitor card is a troubleshooting device all in itself.

The following is an excerpt provided by LiteTouch for troubleshooting the monitor card.

The following questions and suggestions should be used to resolve problems with the Monitor Card:

1) Is it the Monitor Card that's the problem or is it the rest of the system that is the problem? This question can be answered by trying another Monitor Card and seeing if the problem goes away.
2) What part is failing—the module bus, the station bus, or the power?
3) If the module bus is failing, is it on the transmit side, receive side, or both? A transmit-side problem may be due to a bad C2000 Card. A bad receive side may be caused by blown surge protection or bad wiring. If both sides are failing, check the installation of the Monitor Card. Perhaps there is a bad connection.
4) If the station bus is failing, is it on the transmit side, receive side, or both? A transmit-side problem may be due to a bad C2000 Card. A bad receive side may be caused by blown surge protection on the Back Plane Board or bad wiring.
5) If the Power LED doesn't come on and the CPU and C2000 Cards do not run, check the Power Switch, Power Supply cable, and connector for proper voltages; ensure that fuses are intact and ensure the Monitor Card is securely installed in the right-most slot of the Back Plane Board (Source: LiteTouch).

### C2000 Card Troubleshooting

The C2000 card can be installed in any slot of the back plane board except the right-most slot. The function of this card is to interface between the 2000 control stations/modules and the 5000LC CPU card. It can also be used as a backup to the CPU card.

The following excerpt is provided by LiteTouch for troubleshooting the C2000 system card.

The following questions and suggestions should be used to resolve problems with the C2000 Card:

1) Is it the C2000 Card or the rest of the system that is the problem? This question can be answered by trying another C2000 Card and seeing if the problem goes away.

2) What status code (if any) is the C2000 Card displaying? See codes for ideas.
3) Is the shorting bar (JP1) installed on Boot Block Connector (J2)? Failure to install this part will prevent the C2000 from operating.
4) Is the problem on the station bus or the module bus? Is it on only one port or on all ports?
5) Do the Monitor Card LEDs indicate that the C2000 communication signals are getting through?

The following information lists the status indicator codes and their interpretations for the C2000 card:

| 7-Segment Code | Stands For | Meaning |
|---|---|---|
| Rotating Figure-8 Snake | Running | System is running and there are no errors detected |
| Flashing L | Load Mode | C2000 is ready to upload new firmware |
| Flashing C | No Card | CPU Card is not responding. |
| Flashing F | FIFO Full | The C2000 Transmit FIFO is Full. |
| Flashing d | Invalid Mode | C2000 Card is in an invalid mode - not load or operate. |

(Source: LiteTouch)

### *CPU Card Troubleshooting*

The CPU card can be installed in any slot of the back plane board except the right-hand slot. This card contains the job programming and the system clock information. The CPU card basically does the "thinking" for the 5000LC system.

The following is an excerpt provided by LiteTouch for troubleshooting the CPU card.

The following questions and suggestions should be used to resolve problems with the CPU Card:

1) Is it the CPU Card or the rest of the system that is the problem? This question can be answered by trying another CPU Card and seeing if the problem goes away.
2) What status code (if any) is the CPU Card displaying? See Codes for ideas.
3) Is the shorting bar (JP1) installed on the Boot Block Connector (J9)? Failure to install this part will prevent the CPU Card from operating.

4) Was the job programming uploaded?
5) Can you communicate through the serial port with the CPU Card?
6) What are the results of diagnostics?

The following information lists the status indicator codes and their interpretations for the CPU card:

| 7 - Segment Code | Stands For | Meaning |
|---|---|---|
| Rotating Figure-8 Snake | Running | System is running and there are no errors detected. |
| Flashing C | No Card | C2000 card is not responding. |
| Flashing d | Invalid Mode | CPU Card is in an invalid mode - not load or operate. |
| Flashing E | FIFO Full | The Modem Card Receive FIFO is Full. |
| Flashing F | FIFO Full | The C2000 Card Receive FIFO is Full. |
| Flashing J | Job Error | Job programming has invalid CRC. |
| Flashing L | Load Mode | CPU Card is ready to upload new firmware. |

(Source: LiteTouch)

### *Modem Card Troubleshooting*

The modem card is used as the telephone interface for the 5000LC system. It can provide the voice menuing and touch-tone switch control. Job uploading and downloading can be accomplished through the modem card as well as the CP5000 diagnostics.

The following is an excerpt provided by LiteTouch for troubleshooting the modem card.

The following questions and suggestions should be used to resolve problems with the Modem Card:

1) Is it the Modem Card or the rest of the system that is the problem? This question can be answered by trying another Modem Card and seeing if the problem goes away.
2) What status code (if any) is the Modem Card displaying? See codes for ideas.
3) Is the shorting bar (JP1) installed on the Boot Block Connector (J9)? Failure to install this part will prevent the Modem Card from operating.
4) Can you communicate through the serial port with the Modem Card?
5) Does the Modem Card answer an incoming call?

The following information lists the status indicator codes and their interpretations for the modem card:

| 7-Segment Code | Stands For | Meaning |
|---|---|---|
| Rotating Figure-8 Snake | Running | System is running; no errors detected. |
| Flashing C | No Card | CPU Card is not responding. |
| Flashing d | Invalid Mode | Modem Card is in invalid mode - not load or operate. |
| Flashing F | FIFO Full | The Modem Card Transmit FIFO is Full. |
| Flashing L | Load Mode | Modem Card is ready to upload new firmware. |

(Source: LiteTouch)

## Conclusion

As was mentioned earlier, it is nearly impossible to anticipate every troubleshooting situation. Every effort has gone into this troubleshooting chapter to illustrate some of the most common and unusual situations. Even though your specific home automation problems may differ from the ones mentioned here, some of the same troubleshooting principals referred to in this chapter may be used to arrive at your own successful problem-solving conclusions.

When a problem arises that cannot be easily solved, you are not necessarily "stuck" with figuring it out for yourself. LiteTouch operates a technical support line at (1-800-527-5527). These technical support people are there to help you with your specific troubleshooting problems. It kind of goes back to the old saying that two heads are better than one.

# 7

# The Past, Present, and Future of Home Automation

## An Older Home Automation System

The roots of home automation travel back quite a few years. One early type of home automation system consisted of massive banks of low-voltage relays mounted into electrical junction boxes. When this type of system was installed in a larger home, there were literally hundreds of cables and thousands of individual wires running to these junction boxes for home control. The control wires would typically run from these junction boxes to small low-voltage momentary rocker switches located throughout the home. There would be single points of load control and also master control stations. The master stations would control any group of relays that were hardwired into its system. Blocking diodes were commonly used in these master circuits to limit power back-feeds.

When one or more of these blocking diodes would fail, erroneous load activations would occur and it would be up to the technician to find the bad one(s). This job was usually a nightmare and subjected the technician to a snarl of multicolored wires and hundreds of blocking diodes. The relays would sometimes burn out when a momentary switch failed in an on or off position. This required the faulty switch, in addition to the bad relay, to be found. If the relay panels were not marked properly, finding the bad one could turn out to be a real chore. If the bad momentary switch would not show itself because of intermittent switch button sticking, you were faced with a dilemma. Either you replaced all of the switches in the relay circuit that could be causing the problem, or you guessed at which switch buttons were probably sticking and then replaced only those switches. The latter decision increased the risk of further relay failures. While this type of system was generally not very pleasant to work on, the ancestry of modern home automation predicates from these early low-voltage systems.

## Integration and Retrofit Products

Currently, there are many different home automation-type products and systems on the market. The selections range in price all the way from the inexpensive to the expensive. Some manufacturers offer complete home automation control and others offer specific points of control. HVAC, home entertainment, appliance control, and other automated products can be purchased either as an integrated system or strictly as a separate item.

Certain studies show that a majority of people (regardless of age) are interested in home automation in one form or another. Not everyone has the same desire as to what equipment needs to be automated in his or her home, so systems or products that offer a certain degree of flexibility seem to be a preference among homeowners. There is a distinctly separate home automation market that caters to retrofitting a home that already has conventional systems installed. Radio frequency, analog, infrared, and other similar controlling equipment allow retrofit into older homes with a minimum of expense. These systems are generally not as reliable as a digital-type system, but they do serve a definite purpose and have gained some popularity. Some of the products that fit into this category are RF, IR, analog-controlled switches, dimmers, wall outlets, and wireless home entertainment. Of course, the list is considerably larger than this with new products being introduced to the home automation market every day.

## Existing and Future Technology

There are some specialty devices used in home automation that are becoming more reliable and popular. Accurate voice recognition devices were previously too expensive to integrate into a typical home automation environment. Now, however, these items are becoming more reliable and the price for them has dropped considerably. Pressure pad and motion detection are other products that have become more affordable and reliable. Motion detection equipment is sometimes subjected to certain temperature variables that interfere with normal operation. Integrating temperature-compensating circuits into this motion (or heat detection) equipment has made it more reliable and useful. Pressure pads can also be installed in many parts of the home, including heavy traffic areas. The newer pressure pads tend to have very good anti-wear characteristics and can be installed to activate almost any automated system component.

The products that are being developed in the field of home automation seem to be geared toward better interfacing with the human body. Fingerprint scans, face recognition, and devices that change specific human features into a digital type pattern are being researched or developed. Retinal and iris identification scans have been around for several years, but the expense for accurate scanning equipment has been a little high. Touch-screen technology is another item that has been around for some time. Computer screens located throughout the house typically show a floor-plan layout of the home. Pressing their corresponding icons located on the touch-sensitive screen can control entertainment, environment, security, and lighting. This technology has been around longer than you might think, but it has been out of the price range of most people. The future automated home will probably offer all of this special equipment at some lower, more affordable price.

## Home Automation Standards

In all likelihood, home automation will become more common once certain industry-wide standards are implemented and conformed to. Many standards have already been set. Since it is not as common as, say, microwave ovens are in a home, there has not been a great need yet. When home automation compatibility starts becoming an issue, standards will probably be at the forefront of the solution. We can all go down to a computer supply store and find equipment that we need without worrying too much about compatibility. This was not always the case with regard to most computers of the past. Many times you needed to purchase the exact proprietary item if you wanted it to work with your particular computer. Accepted personal computer industry standards are why this convenience is now afforded to us. The day we can walk into a local store and purchase compatible home automation equipment like we now buy a color television, is the day we know that certain standards are accepted industry-wide.

## Conclusion

The diversity of home automation systems currently on the market reveals to us how technology changes as we go about our daily lives. Since industry competition tends to change even the most exclusive automation systems, change for the LiteTouch system is inevitable.

Many of the principles and practices referred to in this book are meant to be helpful for many years to come. With regard to this, learning the system as it is now should ease you into the newer changes as they come about. Considering that so many of these systems are currently in use, servicing and upgrading them should be a very good business opportunity in the future. There always seems to be room for anyone who desires to enter any field, and home automation is no exception.

Like all successful businesses, home automation relies heavily on customer satisfaction. People are generally willing to pay a good price for good service. They usually realize that the home automation system that they are purchasing will have problems on occasion. However, they want to feel secure that their future problems will be taken care of in an honest and competent manner. In many instances, money is not the main issue with homeowners; they just want good reliable service.

Home automation is an exciting and satisfying field of endeavor. It permits our imagination to create a unique living environment that demonstrates the current era of technology that we all live in. Although technology will change as time goes on, the sound principles of working with that technology will not.

# Index

## Symbols



## A



## B



## C

# D

## E



## F



## G



## H

## I



## K



## L



## M

## N



## O



## P



## R

# S

## T



## U

PROMPT®
PUBLICATIONS

## RadioScience Observing
### Volume 1
*by Joseph Carr*

Among the hottest topics right now are those related to radio: radio astronomy, amateur radio, propagation studies, whistler and spheric hunting, searching for solar flares using Very Low Frequency (VLF) radio and related subjects. Author Joseph Carr lists all of these under the term "radioscience observing" — a term he has coined to cover the entire field.

In this book you will find chapters on all of these topics and more. The main focus of the book is for the amateur scientist who has a special interest in radio. It is also designed to appeal to amateur radio enthusiasts, shortwave listeners, scanner band receiver owners and other radio hobbyists.

RadioScience Observing also comes with a CD-ROM containing numerous examples of radio frequencies so you can learn to identify them. It also contains detailed information about the sun, planets and other planetary bodies.

Communications Technology
336 pages • paperback • 7-3/8 x 9-1/4"
ISBN: 0-7906-1127-9 • Sams: 61127
$34.95

## RadioScience Observing
### Volume 2
*by Joseph Carr*

Joe Carr expands on Volume 1 with all-new material, covering: techniques and methods; hardware design and construction; more RadioScience theory; and related geo-science and planetary science activities.

The discussions are more skill-oriented, and the design and construction more specific, building on the material presented in Volume 1. Again, a CD-ROM is included, containing information for making calculations, as well as audio clips, enabling readers to actually experience the thrills of "listening to the heavens."

Chapter topics include: Propagation Anomalies; Electromagnetic Fields and Safety; Using Used TVRO Dishes in RadioScience Observing; Finding Compass Bearings; Seismic Observations; Ultraviolet and Ozone Measurements; Receiver Selection; HF Low-Noise Amplifiers; Radio Telemetry on a Budget; Setting Up for SETI; and much more!

Communications Technology
440 pages • paperback • 7-3/8 x 9-1/4"
ISBN: 0-7906-1172-4 • Sams: 61172
$34.95

**To order today or locate your nearest Prompt® Publications distributor at 1-800-428-7267 or www.samswebsite.com**

Prices subject to change.

PROMPT® PUBLICATIONS

## IC Design Projects

*Stephen Kamichik*

*IC Design Projects* discusses some of the most popular and practical ICs, and offers you some projects in which you can learn to create useful and interesting devices with these ICs.

Once you have read through this book and completed some of its projects, you will have a real, working knowledge of ICs, enabling you to design and build you own projects!

Topics include: how power supplies operate, integrated circuit voltage regulators, TTL logic, CMOS logic, how operational amplifiers work, how phase-locked loops work, and more!

Projects include: battery charger, bipolar power supply, capacitance meter, stereo preamplifier, function generator, DC motor control, automatic light timer, darkroom timer, LM567 tone decoder IC, electronic organ, and more!

Projects
261 pages • paperback • 7-3/8 x 9-1/4"
ISBN: 0-7906-1135-X • Sams 61135
$24.95

## Build Your Own Test Equipment

*Carl J. Bergquist*

Test equipment is among the most important tools that can be used by electronics hobbyists and professionals. Building your own test equipment can carry you a long way toward understanding electronics in general, as well as allowing you to customize the equipment to your actual needs.

*Build Your Own Test Equipment* contains information on how to build several pragmatic testing devices. Each and every device is designed to be highly practical and space conscious, using commonly-available components.

Projects include: Prototype Lab, Multi-Output Power Supply, Signal Generator and Tester, Logic Probe, Transistor Tester, IC Tester, Portable Digital Capacitance Meter, Four-Digit Counter, Digital Multimeter, Digital Function Generator, Eight-Digit Frequency Counter, Solid-State Oscilloscope, and more.

Professional Reference
267 pages • paperback • 7-3/8 x 9-1/4"
ISBN: 0-7906-1130-9 • Sams: 61130
$29.95

**To order today or locate your nearest Prompt® Publications distributor at 1-800-428-7267 or www.samswebsite.com**

Prices subject to change.

PROMPT®
PUBLICATIONS

## Electronic Circuit Guidebook, Volume 1: Sensors

*Joseph Carr*

Many texts discuss instrumentation and computer interfacing. A topic usually lacking in these texts, the analog interface, is covered in *Electronic Circuit Guidebook, Volume 1: Sensors.*

In this volume you will find information you need about typical sensors, along with much more information about analog sensor circuitry. Amplifier circuits are well covered, along with differential amplifiers, analog signal processing circuits, and more.

Topics covered include: Noise, Sensors and Instruments, Instrument Design, Resistive Sensors, Electro-Optical Sensors, Magnetic Sensors, Interfaces, Building the Analog Subsystem, Analog Amps, Inverting and Noninverting Amps, Differential Amps, Analog Signal Processing Circuits, Removing 60 Hz Noise, Shield and Grounding Methods, A/D-D/A Converters, Sensor Resolution Improvement Techniques, DC Supply Circuits, and more.

Electronics Technology
339 pages • paperback • 7-3/8 x 9-1/4"
ISBN: 0-7906-1098-1 • Sams: 61098
$29.95

## Electronic Circuit Guidebook, Volume 2: IC Timers

*Joseph Carr*

Part I of this book demonstrates the theory of how various timers work. This is done by way of an introduction to resistor-capacitor circuits, and in-depth chapters on TTL and CMOS digital IC timers, the LM-555 and other ICs, operational amplifier timer circuits, retriggerable timers and long duration timers. The simplified equations and detailed graphics are perfect for both technicians and hobbyists.

Part II presents a variety of different circuits and projects, some standalone, and others for incorporation into other circuits. Examples included: touch-plate trigger, missing pulse detector, analog audio frequency meter, one-second timer/flasher, relay and optoisolator drivers, two-phase digital clock, 100kHz crystal calibrator and many more. *Electronic Circuit Guidebook, Volume 2: IC Timers* will teach you enough to be able to rework and modify the circuits covered here, and also design a few of your own.

Electronics Technology
239 pages • paperback • 7-3/8 x 9-1/4"
ISBN: 0-7906-1106-6 • Sams: 61106
$29.95

**To order today or locate your nearest Prompt® Publications distributor at 1-800-428-7267 or www.samswebsite.com**

Prices subject to change.

PROMPT®
PUBLICATIONS

## Home Automation Basics

### Practical Applications Using Visual Basic 6

*by Tom Leonik, P.E.*

This book explores the world of Visual Basic 6 programming with respect to real-world interfacing, animation and control on a beginner to intermediate level.

*Home Automation Basics* demonstrates how to interface to a home automation system via the serial port on your PC. Using a programmable logic controller (PLC) as a home monitor, this book walks you through the process of developing the home monitor program using Visual Basic programming. After programming is complete the PLC will monitor the following digital inputs: front and rear doorbell pushbuttons, front and rear door open sensors, HVAC systems, water pumps, mail box, as well as temperature controls. The lessons learned in this book will be invaluable for future serial and animations projects!

Electronics Technology
386 pages • paperback • 7-3/8" x 9-1/4"
ISBN 0-7906-1214-3 • Sams 61214
$34.95

## Howard W. Sams Laser Design Toolkit

*Carl Bergquist*

For over 30 years, the laser has enriched our lives to an extent few can fully appreciate. Until recently, the cost for the hobbyist was far to prohibitive. Now however, this is no longer the case!

There is so much to lasers, that it's hard not to like them once you get to know them. It is impossible to describe the thrill from a red spot on the wall, projected from a helium-neon laser built with your own hands. Like most tubes, they have a life span of several thousand hours, and run at full performance for most of that span.

The *Laser Design Toolkit* introduces you to this fascinating field and helps you complete exciting projects like building your first HeNe laser, working with optics, constructing a laser light show, hologram processing, building a laser light meter, working with laser diodes, and more!

Projects
336 pages • paperback • 7-3/8 x 9-1/4"
ISBN: 0-7906-1183-X • Sams 61183
$29.95

**To order today or locate your nearest Prompt® Publications distributor at 1-800-428-7267 or www.samswebsite.com**

Prices subject to change.

PROMPT®
PUBLICATIONS

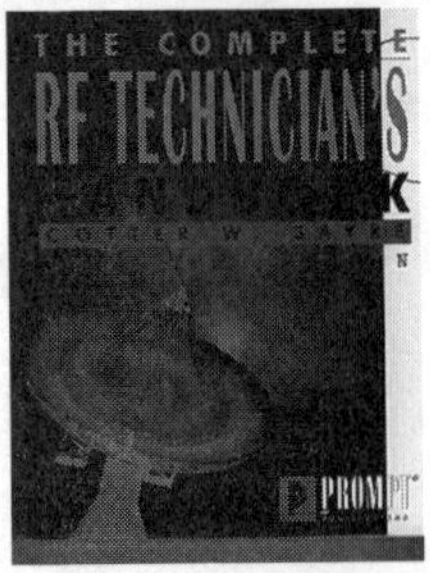

## The Complete RF Technician's Handbook, *Second Edition*

*Cotter W. Sayre*

This is THE handbook for the RF or wireless communications beginner, student, experienced technician or ham radio operator, furnishing valuable information on the fundamental and advanced concepts of RF wireless communications.

Circuits found in the majority of modern RF devices are covered, along with current test equipment and their applications. Troubleshooting down to the component level is heavily stressed. RF voltage and power amplifiers, classes of operation, configurations, biasing, coupling techniques, frequency limitations, noise and distortion are all included, as well as LC, crystal and RC oscillators. RF modulation and detection methods are explained in detail – AM, FM, PM and SSB – and associated circuits, such as AGC, squelch, selective calling, frequency multipliers, speech processing, mixers, power supplies, AFC, multiplex-

Communications Technology
366 pages • paperback • 8-1/2 x 11"
ISBN: 0-7906-1147-3 • Sams: 61147
$34.95

## The Right Antenna *Second Edition*

*Alvis J. Evans*

Television, FM, CB, cellular phone, satellite and shortwave signals are available in the air to anyone, but it takes a properly selected and installed antenna to make them useful. With easy-to-understand text and clearly illustrated examples, *The Right Antenna* will give you the confidence to select and install the antenna that meets your needs.

*The Right Antenna* explains how antennas work and breaks them into TV and FM for discussion. A separate chapter is devoted to interference, and antennas used by hams for antenna band operation. This new second edition also includes chapters on DSS, other satellite antennas, and TV and amplifier antennas.

Also included: How Antennas Work, Selection of Antennas for Specific Needs, Installation of Antennas, Fringe Area and MATV Antennas, How to Identify and Eliminate TV Interference.

Communications Technology
120 pages • paperback • 6 x 9"
ISBN: 0-7906-1152-X • Sams: 61152
$24.95

**To order today or locate your nearest Prompt® Publications distributor at 1-800-428-7267 or www.samswebsite.com**

Prices subject to change.

PROMPT® PUBLICATIONS

## Speakers For Your Home and Automobile, *Second Edition*

*Gordon McComb, Alvis Evans, Eric Evans*

The cleanest CD sound or the clearest FM signal are useless without a fine speaker system. *Speakers for Your Home and Automobile* will show you the hows and whys of building quality speaker systems for home or auto installation. With easy-to-understand instructions and clearly-illustrated examples, this book is a must-have for anyone interested in improving their sound systems.

The comprehensive coverage includes:

Speakers, Enclosures, and Finishing Touches
Construction Techniques
Wiring Speakers
Automotive Sound Systems and Installation
Home Theater Applications

Audio
192 pages • paperback • 6 x 9"
ISBN: 0-7906-1119-8 • Sams: 61119
$24.95

## Complete Guide to Audio

*John J. Adams*

*Complete Guide to Audio* was written for the consumer who wants to know more about sound systems. With comprehensive, simple explanations, it answers questions you may have asked salespeople in the past but were unable to get answers for.

In addition, this book explains some common problems that you may run into while setting up your home entertainment center. The information in the book will help you make successful purchasing decisions and demystify the jungle of wires and connections that come with your audio system.

Topics include: Audio Basics, Sound, Stereo-phonics, Home Theater, Amplifiers and Preamplifiers, Receivers and Surround-Sound, Cassette and CD Decks, DVD, MiniDisc and Phonographs, Speakers, Headphones and Microphones, Computer Sound, Brands and Choices, Hookups and Accessories.

Audio
163 pages • paperback • 7-1/4 x 9-3/8"
ISBN: 0-7906-1128-7 • Sams: 61128
$29.95

**To order today or locate your nearest Prompt® Publications distributor at 1-800-428-7267 or www.samswebsite.com**

Prices subject to change.

PROMPT®
PUBLICATIONS

## Optoelectronics, *Volume 3*

*Vaughn D. Martin*

Here in *Optoelectronics, Vol. 3* is a lab manual designed to teach you how to design and build optoelectronics devices by assembling the projects.

Volume 3 gives you the necessary instructions and training to design and assemble your own optoelectronic devices, including a laser optics lab, a sandbox holography lab, and a physical optics lab. It also offers you a review of the essentials of optoelectronics covered in the previous volumes.

Topics include: Light's Interaction with Surfaces and Optical Elements, Optoelectronic Measurements, Measuring Instruments and Projects, Light and Laser Sources, Scope Cameras, Edge Detection ICs, Reflections and Mirrors, Speed-of-Light Experiment, Refraction, Polarization, Dispersion, Internal Reflections and Photo-elasticity, Ambient Light Suppression Techniques, Optical Spectrum Analyzer, Optical Time Domain Reflectometer, AND MORE!

Professional Reference
435 pages • paperback • 8-1/2 x 11"
ISBN: 0-7906-1122-8 • Sams: 61122
$44.95

## Communication Electronics for Technicians

*Joseph Carr*

In the *Communication Electronics for Technicians*, noted and respected author Joseph Carr explains why the CET is an important and valuable addition to your professional resume. Mr. Carr, who has been CET certified since the early '70s, has composed a comprehensive text in three sections, divided into Associated Level, Journeyman Level, and a "Body of Knowledge".

The first two sections contain a question-answer-discussion chapter, so that you can not only see the correct answer, but also the reasoning behind it. Following that, a practice examination chapter tests your knowledge in a multiple-choice format with answer key. The third section represents a set of facts every journeyman in this trade needs to know, especially one who is thinking about attaining Master Technician status.

Professional Reference
320 pages • paperback • 7-3/8 x 9-1/4"
ISBN: 0-7906-1156-2 • Sams: 61156
$39.95

**To order today or locate your nearest Prompt® Publications distributor at 1-800-428-7267 or www.samswebsite.com**

Prices subject to change.

PROMPT® PUBLICATIONS

## The Microcontroller Beginner's Handbook, *Second Edition*

*Lawrence A Duarte*

*The Microcontroller Beginner's Handbook* examines all elements of microcontroller use, including such industrial considerations as price vs. performance and firmware. A wide variety of third-party development tools is also covered, both hardware and software, with emphasis placed on new project design. There are many ideas and hints for an experienced microcontroller user. When done, the reader will be able to design, write, and build a new project.

Topics include: What is a Microcontroller? Binary and Boolean Logic; Bits & Bytes; Analog to Digital—Digital to Analog; The PIC Family of Microcontrollers; PIC Firmware; Project: The RS232 Terminal; AC Motor Control; A Telephone Busy Buster; A Video Output Thermometer; and more.

Includes CD-ROM with source code examples, sample assemblers, and over 4,500 data sheet pages.

Electronics Technology
256 pages • paperback • 7-3/8 x 9-1/4"
ISBN: 0-7906-1153-8 • Sams 61153
$24.95

## Power Supply Troubleshooting & Repair

*Lanny L. Logan*

*Power Supply Troubleshooting & Repair* is designed to provide technicians with a better understanding of how switched-mode power supplies operate. It also provides practical, useful procedures to follow when troubleshooting switched-mode power supplies.

Focusing on the most common supplies found throughout the industry, using block diagrams and a full-circuit schematics, this book presents various circuits throughout its chapters with brief troubleshooting procedure for servicing each supply.

Topics covered include:Switched-mode power supply basics; Pulse width-controlled supplies; Free-running supplies; Discrete switch transistor supplies; IC-controlled supplies; Fixed-frequency supplies; Linear supply troubleshooting; Discrete shutdown circuits; Built-in switches; AND MORE!

Troubleshooting and Repair
205 pages • paperback • 7-3/8 x 9-1/4"
ISBN: 0-7906-1138-4 • Sams 61138
$24.95

**To order today or locate your nearest Prompt® Publications distributor at 1-800-428-7267 or www.samswebsite.com**

Prices subject to change.

PROMPT®
PUBLICATIONS

## Audio Systems Technology, Level I

### *Handbook for Installers and Engineers*

*NSCA*

This book is a one-stop information source for today's audio technician. It can be used as a study guide to prepare for NICET audio technician certification exams, as well as a comprehensive reference for the installer of audio systems—both out in the field and at the bench.

Designed to correspond with Level I work elements on the NICET tests, this valuable handbook presents the basics of audio installation as it is practiced in the industry today. Topics include: Basic Electronic Circuits; Basic Math; Basics of Microphones & Loudspeakers; Basic Wiring; Switches and Connectors; Codes, Standards and Safety; and Reading Plans and Specifications.

Additionally, information about getting certified by NICET is included, with tips and strategies to help with your success on the NICET exams.

Audio
320 pages • paperback • 7-3/8 x 9-1/4"
ISBN: 0-7906-1162-7 • Sams 61162
$34.95

## Audio Systems Technology, Level II

### *Handbook for Installers and Engineers*

*NSCA*

Designed to correspond with Level II work elements on the NICET tests, this book presents intermediate level content on audio installation as it is practiced in the industry today. Some of the topics: Audio Calculations; Acoustical Measurements; Microphones, Loudspeakers and Mixers; Trigonometry and Geometry; Wiring and Cabling; Effective Business Communication; and Bench Test Equipment.

Information about getting certified by NICET is included, with tips and strategies for the test-taker.

Both Level I and Level II books are for you if:

- You want to be NICET-certified in audio systems.
- You are a developing audio installer or engineer.
- You want to enhance your knowledge of audio systems design and installation.
- You are experienced but want to brush up on intermediate-level audio.

Audio
432 pages • paperback • 7-3/8 x 9-1/4"
ISBN: 0-7906-1163-5 • Sams 61163
$39.95

**To order today or locate your nearest Prompt® Publications distributor at 1-800-428-7267 or www.samswebsite.com**

Prices subject to change.